WAVERLEY
Carlisle · Hawick · Galashiels · Edinburgh

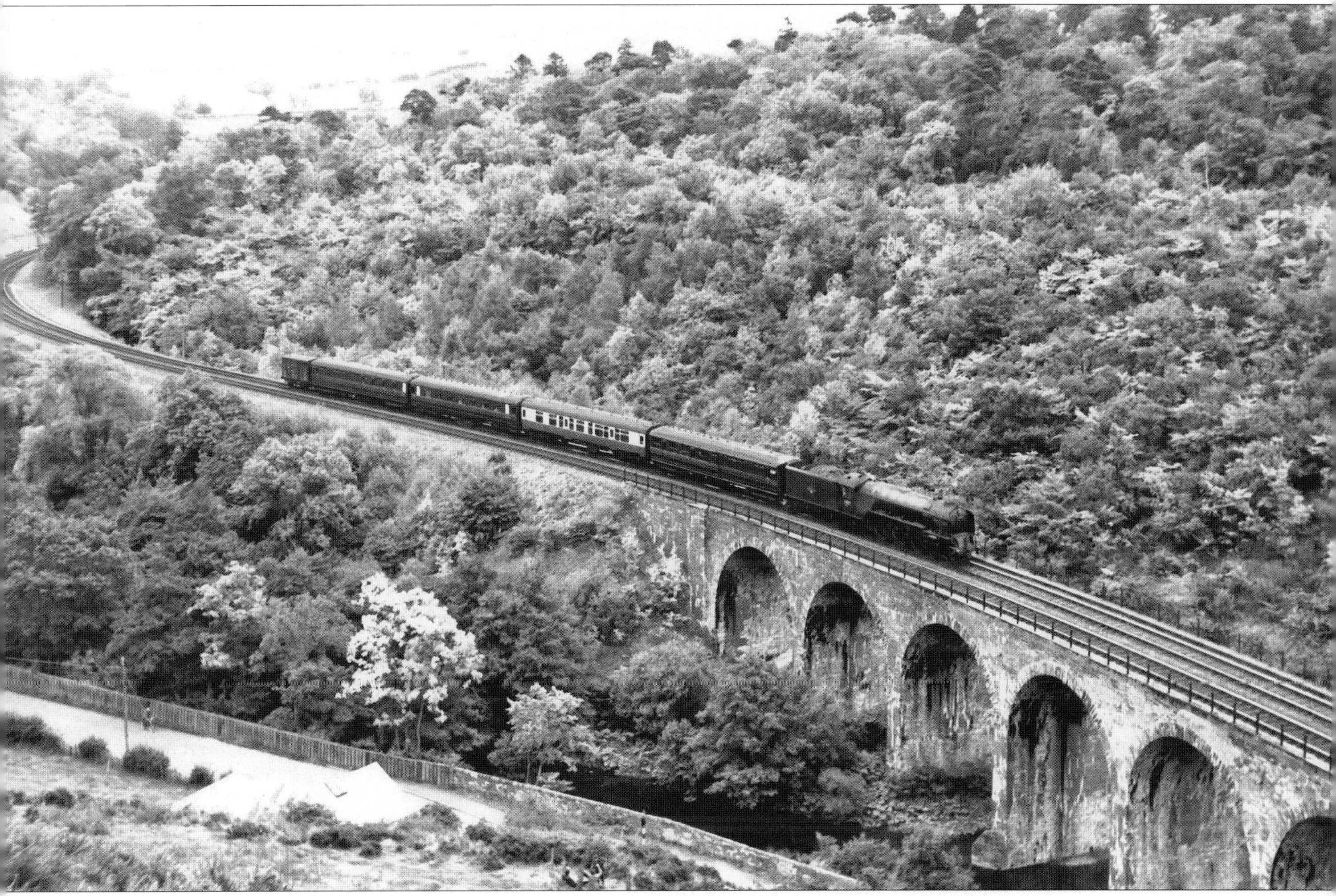

Images from the Transport Treasury archive

Compiled by Jeffery Grayer

Reviving the memories of yesterday…

© Images and design: The Transport Treasury 2024, Text: Jeffery Grayer.
ISBN 978-1-913251-79-6
First published in 2024 by Transport Treasury Publishing Ltd., 16 Highworth Close, High Wycombe, HP13 7PJ
www.ttpublishing.co.uk

Printed by Short Run Press Ltd., Exeter.

The copyright holders hereby give notice that all rights to this work are reserved.
Aside from brief passages for the purpose of review, no part of this work may be reproduced, copied by electronic or other means, or otherwise stored in any information storage and retrieval system without written permission from the Publisher. This includes the illustrations herein which shall remain the copyright of the copyright holder.

Front Cover : Two small children on Hawick's down platform are more interested in looking at the River Teviot flowing beneath part of the station platforms than witnessing the arrival of Edinburgh Haymarket's Class A3 No. 60057 *Ormonde* with an up service in 1954. Like many of its classmates this example was named after a racehorse, in this case the winner of not only the Derby in 1886, but also the 2000 Guineas and the St. Leger. *(Neville Stead Collection)*

Frontispiece : Epitomising the dramatic scenery through which much of the Waverley route ran is this view of A1 Class Pacific No. 60159 *Bonnie Dundee* traversing Slitrig, also known as Lynnwood, viaduct consisting of six arches which crossed over Slitrig Water and the B6399 road to the south of Hawick with the 3.22pm service from Carlisle to Edinburgh on 26 September 1960. The Pacific's name, formerly carried by Reid D29 class 4-4-0 No. 62413 which had been scrapped in August 1950, was appropriated for the A1 in November of that year. The viaduct was demolished in October 1982 so if the line is ever reinstated south of Hawick some expensive reconstruction will be required.
(W. A. C. Smith)

Rear Cover : Negotiating some of the typical bleak Border country traversed by the Waverley route on 26 September 1964 is Class A3 No. 60085 *Manna* with one of the frequent Carlisle – Millerhill yard freight services. Just a fortnight later this fine Gresley Pacific would be withdrawn from service at Gateshead shed and summarily scrapped by Draper's at their yard in Hull in January 1965. *(W. A. C. Smith)*

Our sincere thanks go to the following people who contributed to the accuracy of this book:

Malcolm Wells	Mike Yeoman
David Alexander	Robert Day
Alasdair Taylor	Steve Andrews
John Kilford	Roger Bates
David Percival	Bob Allen
Jack Kernahan	Geoffrey Smith
Bill Armstrong	Andrew Morrey
John Kalaitzis	

CONTENTS	PAGE
Introduction	4
Route Map	5
Carlisle – Riccarton Junction	6

Carlisle Citadel – Lyneside – Longtown - Riddings Junction – Penton – Kershopefoot – Newcastleton - Steele Road - Riccarton Junction

Whitrope - Galashiels	42

Whitrope – Shankend – Stobs – Hawick – Hassendean – Belses – St. Boswells - Melrose – Galashiels

Stow - Edinburgh	91

Stow – Fountainhall – Heriot – Falahill – Tynehead – Gorebridge – Newtongrange – Hardengreen Junction - Eskbank & Dalkeith – Millerhill – Portobello – Edinburgh Waverley

INTRODUCTION

'The title of this work has not been chosen without the grave and solid deliberation which matters of importance demand from the prudent.'
(WAVERLEY or 'TIS SIXTY YEARS SINCE published 1814)

Thus begins chapter one of 'Waverley' which would give its name to the many novels written by Sir Walter Scott during the period 1814-31. As Scott did not publicly acknowledge authorship until 1827 his series took its name from the title of the first novel with subsequent works bearing the words 'By the author of Waverley' on their title pages. It is perhaps unsurprising that the railway line between Carlisle and Edinburgh through the Scottish Borders should be accorded the title of the 'Waverley Line' given that it passed within a few miles of Scott's baronial mansion built on the proceeds of his writing at Abbotsford near Melrose. The appellation 'Waverley Route' first appeared in North British Railway (NBR) minute books dated 1862 and it was used at the head of the first timetable of Hawick-Carlisle services. The name Waverley has of course also been appropriated for Edinburgh's great railway station and adjacent road bridge and steps and for many other locations throughout the world from North America to Australasia. Additionally it was borne by Thompson Class A2/1 pacific No. 60509 and not forgetting the world's only ocean going paddle steamer 'P.S. Waverley'.

Its fate identified in the Beeching Report of 1963, it was to be almost another 6 years before the short sighted closure of this 98¼ mile line occurred in January 1969, leading to the isolation of much of the Borders region leaving Galashiels and Hawick further from the rail network than any other town of comparable size in the UK. This isolation has been only partially addressed in recent years by the construction of the Borders Railway from Edinburgh through Galashiels to Tweedbank along the course of the old Waverley Line. Given the success of this re-opening in September 2015, traffic having exceeded forecast expectations by some 22% in the first six months alone, the future looks bright for a further extension to Melrose, St. Boswells and Hawick and, perhaps in the long term, re-opening as far as Carlisle should not be discounted. It has been shown time and again that traffic forecasting is very much an inexact science as demonstrated by the fact that three of the new Borders Railway stations, Stow, Galashiels and Tweedbank generated respectively 313%, 330% and a staggering 681% more passengers than had been predicted. I made my own first journey on the new line in August 2021 and I can certainly vouch for the popularity of the trains. Alighting at the current terminus of Tweedbank, a short walk along the banks of the River Tweed brought me to Melrose and the splendid old station where hopefully in the not too distant future, trains may once again be seen. In 2019/20 it was reported by the Office of Rail and Road, formerly the Office of Rail Regulation, that the number of passenger journeys on the route had topped 2m p.a. for the first time and although numbers followed the national trend and plummeted during the Covid pandemic there are signs that they are slowly recovering.

In this volume we feature the majority of the stations on the route together with a variety of motive power that worked the line during the 1950s and 1960s featuring the images of photographers such as Neville Stead, Norris Forrest and W.A.C. Smith whose collections are now in the safekeeping of the Transport Treasury. We begin traditionally at Carlisle's Citadel station and head north through Newcastleton, Hawick, St. Boswells, Melrose and Galashiels finishing at Edinburgh's Waverley station.

Jeffery Grayer
Devon 2024

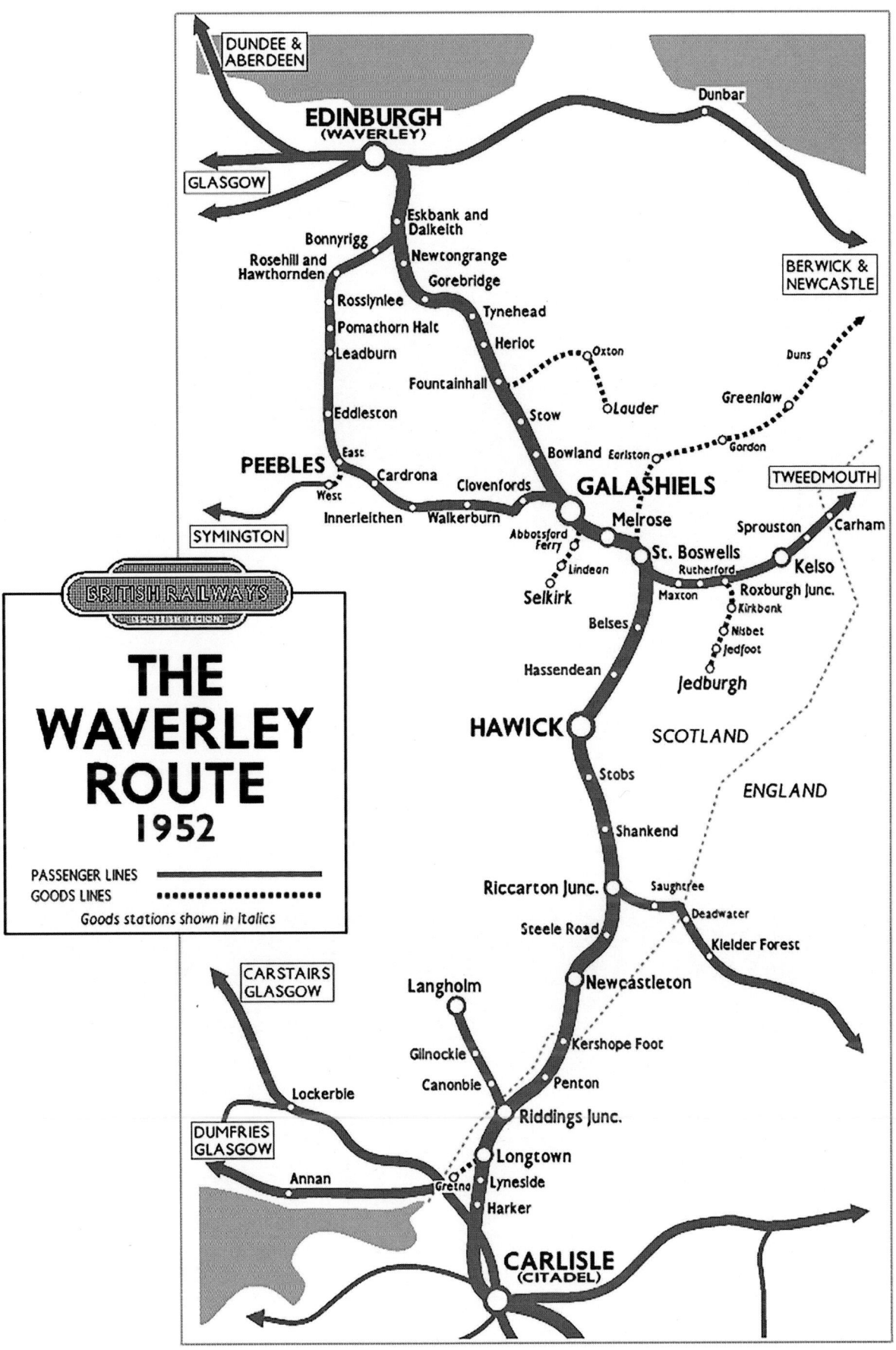

CARLISLE CITADEL

Carlisle - Hawick - Galashiels - Edinburgh

Top left. Handling the service, which used to be known as the 'Thames - Forth' express, is A3 No. 60101 *Cicero* seen at Carlisle awaiting departure for Edinburgh on 1st. September 1954. This view also gives us the opportunity to see the substantial glazed screens that were provided to the large overall roof. Maintenance of this glazed roof was severely neglected during WW2 and in the mid 1950s work began to cut it back and to replace the glazed end screens with more modern, though undoubtedly less attractive, material. *(W. A. C. Smith)*

Bottom left. Continuing the Pacific theme, although this time with the more modern Peppercorn A2 type, is this view of No. 60537 *Bachelor's Button* waiting to leave Carlisle with a Waverley line local service on 2nd. June 1958. Although only built ten years before this image was taken this locomotive would go on to have a very short working life of just 14 years being withdrawn at the end of 1962 – no doubt a victim of dieselisation. *(Sandy McBlain)*

Above. This undated view as well as showcasing A2/1 No.60510 *Robert the Bruce* reveals the extent to which the overall roof of the station had been cut back and the replacement of the glazed end screen. No. 60510 was one of the quartet of A2s with the sub classification A2/1 having been introduced in 1944 as a development of Thompson's original A2/2 Pacifics and incorporating a V2 2-6-2 boiler. All four members of this subclass were withdrawn in 1960/61 and whilst performing adequately were considered to be under boilered and lacking in adhesion. *(Neville Stead Collection)*

Above. D30 Class No. 62425 *Ellangowan* arrives at Carlisle with a Waverley route local from Hawick. Built in 1914 this 'Scott' class 4-4-0 would be withdrawn in 1958 after many years operating over this route much of the time based at Hawick depot. Ellangowan was a fictitious castle featuring in Scott's second Waverley novel 'Guy Mannering'.
(Lens of Sutton)

Right. On 12 July 1961 Haymarket based A2 No. 60535 *Hornet's Beauty* sets out from Carlisle with a service for Edinburgh. It would be reallocated to St. Margaret's depot in October that year and eventually sold to Motherwell Machinery & Scrap Co. in July 1965 after just 17 years of service, the majority being spent in Scotland.
(Sandy McBlain)

Carlisle - Hawick - Galashiels - Edinburgh

WAVERLEY

Top left. Rather overshadowed by the departing down 'Royal Scot', with Coronation Class No. 46221 *Queen Elizabeth* at its head, a Waverley route train waits to leave Carlisle on 17 June 1959. In the adjacent platform a DMU sporting 'speed whiskers' can also be seen. The Carlisle area had been the first on the London Midland Region to see the introduction of DMUs five years previously in 1954 with the introduction of Derby Lightweight sets to the west Cumberland area. *(Sandy McBlain)*

Bottom left. Hawick based Ivatt 2-6-0 No. 43141 leaves Carlisle with the two coach 6.13pm all stations except Steele Road train to Hawick on 23rd. May 1960 scheduled to take 90 minutes for the 45½ mile journey. This formed a useful service for home going shoppers and commuters as the later service on Mondays-Fridays, the last departure of the evening at 7.44 pm, only served the more important stations of Longtown, Newcastleton and Riccarton Junction en route to Hawick and Edinburgh. There were however later departures from Carlisle on Saturdays only serving stations to Langholm and all stations to Hawick with the exception of Stobs. Later that year the 4MT would be transferred from Hawick to Glasgow's Parkhead depot. *(W. A. C. Smith)*

Above. This rather unusual combination sees Ivatt 2-6-0 No. 43139 hauling a couple of DMU cars, Nos. M79126 and M79678, constituting the 4.35pm workmen's train from Parkhouse Halt to Carlisle. It is seen here approaching Dentonholme North Junction on 8 September 1967. Parkhouse Halt, opening in July 1941, was constructed solely for wartime staff working at the nearby RAF 14 Maintenance Unit depot and was not available to the general public. The unit acted as a massive store for the RAF housing everything from aircraft parts, engines and tools to office furniture, stationery and uniforms. Parkhouse Halt closed in January 1969 with closure of the RAF depot, which had its own internal rail network worked by a fleet of diesel shunters, following in September 1996. A single line still exists to the site's sidings which now serve Kingstown Industrial Estate. *(Lens of Sutton)*

WAVERLEY

LYNESIDE

LONGTOWN

Top left. Lacking any suitable sizeable settlement in the vicinity it was decided originally to name the station Westlinton after a small hamlet a short distance away on the main A7 trunk road. Some nine years later it was renamed Lineside although the following year the spelling was changed to Lyneside to reflect the fact that it was situated near the River Lyne at this point. This river, which has its origins in Kershope Forest, eventually flows into the Border Esk river. This image dating from 15 September 1953 is looking north showing the main station building, which today remains in situ as an attractive house, and the level crossing gates which were controlled from the adjacent signalbox.
(Neville Stead Collection)

Bottom left. Closing to passengers as early as November 1929 the station was never well patronised and it can be seen from this view looking south that only very short platforms were provided. It became an unstaffed public siding in June 1956 with closure to goods coming in October 1964. The signalbox remained open for a while after passenger services on the Waverley route were withdrawn in January 1969 as freight was still handled on the section from Carlisle to Longtown, however the box eventually closed in August of that year. *(Neville Stead Collection)*

Above. Longtown was one of the larger settlements between Carlisle and Hawick and in this view looking north taken on 6 March 1960 Gateshead based A1 Pacific No. 60132 *Marmion* heads south with a short freight service. A couple of months later the locomotive would be reallocated to Heaton depot and later in that year would enter Doncaster to have AWS apparatus fitted. Freight was an important traffic over the Waverley route and, as mentioned previously, freight continued to be handled from Longtown after passenger services were withdrawn, this traffic lasting until 31st. August 1970 although a private siding did continue in use after this date. After passenger closure the line south was singled in 1969 and the signalbox reduced in status to that of a gate box. *(Mike Mitchell)*

Above. This view of Longtown looking south shows the signalbox which controlled not only the level crossing but also the junction of the former route to Gretna which branched off to the right immediately after the crossing and which closed to passengers in 1915. After leaving the station the mainline to Carlisle crossed the River Esk on a substantial viaduct which, together with the station, level crossing and junction have all subsequently been demolished. *(Henry Priestley)*

Top right. A southbound freight headed by Class B1 No. 61099 crosses a northbound passenger working on 6 April 1963. Longtown benefitted from the passage of Langholm – Carlisle trains which, in 1963 for example, had the effect of raising the number of services to Carlisle from just four provided by Waverley route trains to ten on weekdays. The subsequent withdrawal of Langholm branch services the following year meant that rail travel from Longtown into the city became considerably less attractive and by 1965/66 there were no trains between the hours of 9.49am and 5.29pm. *(W. A. C. Smith)*

Bottom right. A member of station staff makes his way down the up platform possibly to chastise the photographer who appears to have wandered off the end of the down platform and taken up a stance on the level crossing in order to take his shot on this unrecorded date. However, this view does show the rather ornate water tank mounted on the up platform and in the distance can be seen the goods shed served by sidings on both sides of the line here, additional sidings having been laid during WW2. At one time there was even a small two road locomotive shed located at Longtown but this had closed back in the 1920s. *(Neville Stead Collection)*

Carlisle - Hawick - Galashiels - Edinburgh

RIDDINGS JUNCTION

Top left. An up freight train trundles through Riddings Junction on 14 March 1964 headed by St. Margaret's depot based A1 Pacific No. 60152 *Holyrood*. This station was purely an interchange point for the seven mile Langholm branch which curved away to the left passing over the Liddel Water on a nine arch viaduct thereby crossing the border and entering Scotland. At the time of this view the branch would only have another 3 months of life closing to passengers in June of that year although freight continued to be carried until September 1967. *(Larry Fullwood)*

Bottom left. This unusual view of the station entrance stairs and footbridge reveals that the station building was on two levels with the Booking Office on the upper level – hence the need for stairs. The majority of passengers using the station were no doubt changing trains from the branch to the mainline or vice versa but any local passengers living in the scattered remote farms and cottages in the vicinity wishing to access the platforms had to cross the three goods sidings by means of barrow crossings to gain the stairs with the wall hung lamp seen on the right no doubt guiding their way during the hours of darkness. *(Norris Forrest)*

Above. This view of the down platform shows the sizeable running in board proclaiming that this was the junction for 'Canonbie, Gilnockie and Langholm'. Canonbie boasted a rural colliery although the grade of coal was of a low quality whilst Langholm had several woollen mills. A double sided waiting shelter was provided on the down platform. *(Norris Forrest)*

Ivatt Class 4 2-6-0s were the usual motive power for the branch in the latter years and here Carlisle Kingmoor based No. 43103 has brought the 3.28pm service from Langholm into Riddings Junction on 14 March 1964. It will continue onward to Longtown and Carlisle being due away from the junction after a two minute stop at 3.50pm. This service ran two minutes later on Fridays and Saturdays but still reached Carlisle at the same time every weekday, namely 4.21pm. This 21 mile journey with 4 intermediate stops took 53 minutes at an average speed of 18½ mph. Two coaches comprised the usual loading of branch services and this together with the number of railway staff in evidence can have done little to help the balance sheet of the line. In fact the average number of passengers for each of the 32 weekday trains provided in later years was just 11. *(Larry Fullwood)*

B1 Class No. 61242 passes a platelayers' hut and enters a short cutting with an up freight to the north of Riddings Junction on 6 April 1963. These Thompson 4-6-0s were often to be seen on Waverley line secondary passenger workings between Edinburgh and Galashiels and between Hawick and Carlisle as well as handling freight services and they proved to be very successful in Scotland. This example, named *Alexander Reith Gray*, was one of the batch named after directors of the LNER. *(W. A. C. Smith)*

Although the glory days of St. Margaret's based A3 Pacific No. 60037 *Hyperion* are well and truly behind it and it has been reduced to hauling a down freight, seen here near Riddings Junction on 6 April 1963, it still makes a fine sight. However, it was not to last in traffic much longer being withdrawn at the end of the year after almost 30 years in service. Judging by the sleepers lying at the lineside some renewals are taking place. *(W. A. C. Smith)*

PENTON

Above. Another rather isolated outpost, in which the Waverley line seemed to specialise, was Penton seen here looking north in 1955. In the absence of any local settlement it took its name from nearby Penton House. The signalbox located at the far end of the up platform closed in February 1968 and in the absence of a footbridge the board crossing seen in the foreground was used by staff and no doubt the odd passenger. Like many of the other smaller stations towards the end of services Penton saw only three passenger trains daily in each direction. The small goods yard, which closed in October 1967, was located behind the main station building and notice the check rail on the inside of the up track which negotiated quite a severe curve here. *(Neville Stead Collection)*

Top right. DMUs were introduced to the route in 1966 and this view taken from the front cab of just such a unit shows a local service to Hawick about to leave Penton where a lone railwayman is waiting to board. Even the introduction of these units did not make much appreciable difference to passenger usage and hence revenue, although they certainly enhanced the travelling experience by allowing vision forward over this scenic route. *(Henry Priestley)*

Bottom right. This view affords a closer look at the down platform waiting shelter with its ornate lamp being passed by two railway enthusiasts, and reveals that the signalbox, seen in the previous image, had been reduced to just its brick base having closed in 1968 following the goods yard closure in 1967. The main station building remains today in use as a private house. *(Norris Forrest)*

Carlisle - Hawick - Galashiels - Edinburgh

KERSHOPEFOOT

Top left. This undated image of Kershopefoot station looks north towards the Scottish border which was located immediately after the level crossing as the line crossed the Kershope Burn. The traditional border signs were located at each end of the bridge. There was a small hamlet to the east of the line and some of the trees of the extensive Kershope Forest can be seen in the background to this view. The signalbox which had originally been on the up platform burnt down and was replaced by one on the down platform in 1915 remaining open until the end of services although the station had lost its staffing in March 1967 whilst the goods yard had closed in December 1964.
(Norris Forrest)

Bottom left. This 1955 view shows the whole station with what appears to be a ground shunting signal at the end of the down platform. There were wooden shelters on both platforms whilst off to the right situated by the road were the Station Master's house and adjacent booking office, just about the only reminders of the railway which survive today in residential use. The points in the left foreground led off to a small goods yard consisting of two sidings with freight trains from the south having to reverse to gain access into the yard. *(Neville Stead Collection)*

Above. On 18 April 1965 A4 class No. 60031 *Golden Plover* was captured at Kershopefoot with the 'Scottish Rambler No. 4' railtour which ran from Glasgow to Edinburgh, thence to Carlisle before returning to Glasgow. Judging by the photographers present the train seems to have halted for an unscheduled photographic stop before proceeding to Carlisle. With only five coaches in tow this tour should have presented no problem for the Gresley Pacific even over the challenging Waverley line. *(Neville Stead Collection)*

NEWCASTLETON

Carlisle - Hawick - Galashiels - Edinburgh

Top left. Newcastleton was the first settlement of any relative size north from Carlisle to be served by the Waverley route but even today the population is less than 2000. This view looking north reveals the infamous level crossing gates scene of the protest by several hundred people led by the local minister Rev. Brydon Maben at the closure of the line on Sunday 5 January 1969. The clergyman was arrested by police and following intervention by Liberal MP David Steel who was travelling on the final southbound departure he was released and the sleeper train hauled by D60 *Lytham St. Annes* permitted to proceed. Today the Station Master's house on the far left is all that remains at the site of the station. *(Norris Forrest)*

Bottom left. This view, also looking north, was taken from the footbridge and shows the small station building on the up platform and beyond the access to the goods yard containing five sidings the site of which now does duty as a caravan park. Formerly timber was an important freight traffic here right up until closure of the goods yard in October 1967. There were a number of forested areas in the vicinity including Newcastleton and Tinnisburn Forests. On the platforms can be seen LNER style electric lamp standards. *(Norris Forrest)*

Above. This southwards facing view permits a closer look at the attractive metal lattice footbridge. A short film of a train arriving and departing from Newcastleton on a very inclement day in 1960 is available on the National Library of Scotland website. Newcastleton's relative importance is apparent from the fact that 'The Waverley' express deigned to stop here allowing passengers to be in Carlisle in 32 minutes and Edinburgh in a shade under 2 hours in 1965 for example. Today a bus service leaves the village for Carlisle twice a day and for Hawick three times a day where there is a connection for Edinburgh. *(Norris Forrest)*

WAVERLEY

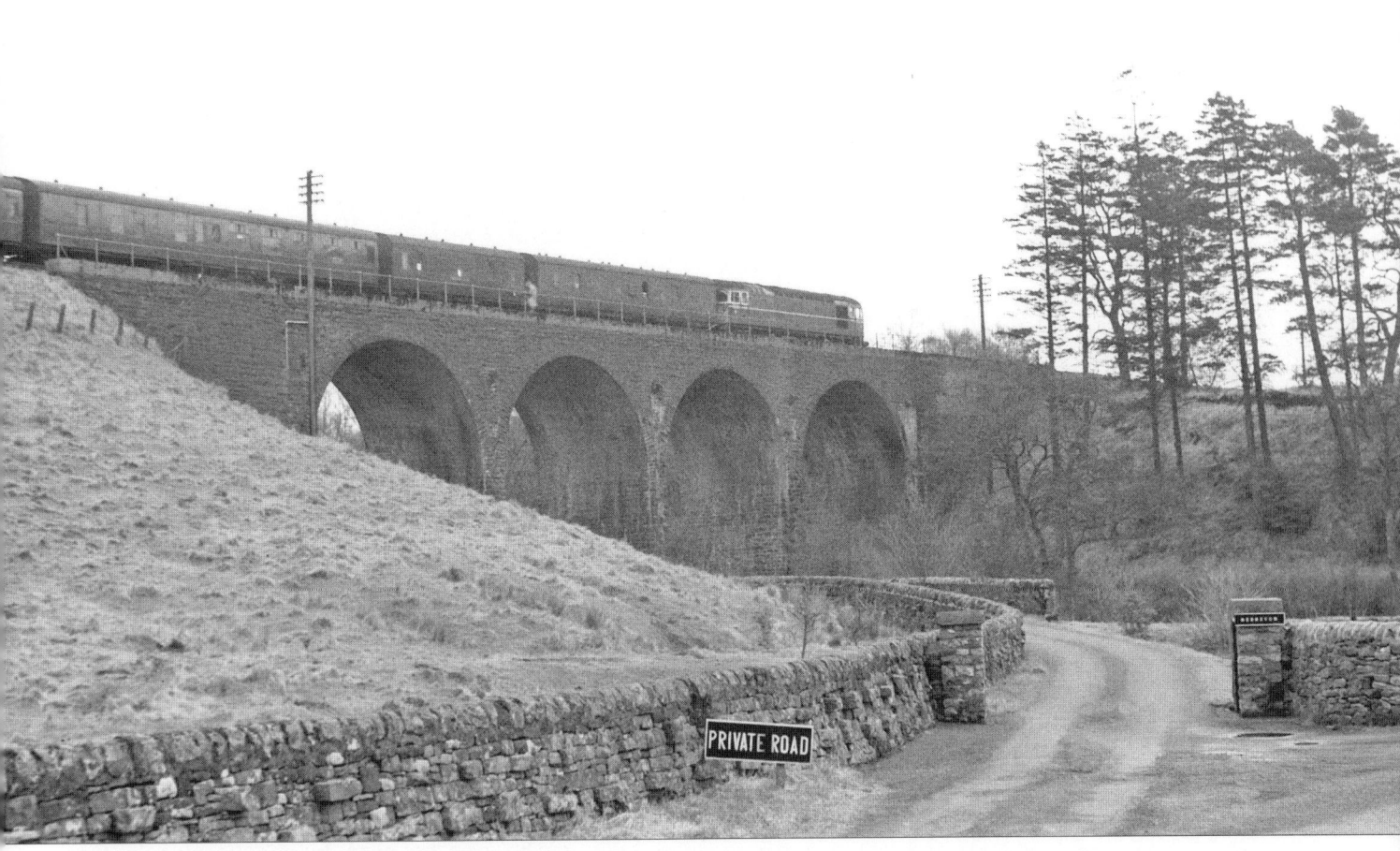

Top left. Our final view of Newcastleton shows in more detail the entrance to the goods yard with its stone built goods shed which had canopies on both sides and a small loading platform on the rail side. A 2 ton crane was also provided for handling the timber traffic. One of the benefits bolstering the case for extending the current line south of Tweedbank to Carlisle is to transfer timber traffic from road reverting to rail haulage. Note the tall up signal with its attractive finial and repeater arm. *(Norris Forrest)*

Bottom left. Introduced in 1936 Gresley's V2 class could be seen in use on the Waverley route for close on 30 years. In this view taken on 5 March 1960 near Newcastleton No. 60953, at that time allocated to Edinburgh's St.Margaret's depot, has charge of an up freight. It would last in service until withdrawal in May 1962. V2s shared the many, and often heavy, freight workings over the route with K3s and A3s and following the opening of the marshalling yards at Carlisle Kingmoor and Edinburgh Millerhill, both of which proved to be very short lived, freight workings were often run hourly throughout the whole 24 hour period. *(Mike Mitchell)*

Above. A couple of miles north of Newcastleton lies the four arch Sandholm or Hermitage viaduct crossing Hermitage Water near to its confluence with Liddel Water. As indicated by the notice a private road serving a dwelling known as Redheugh passed under the nearside arch of the viaduct whilst on the opposite bank the B6399 passed under the viaduct heading towards Hawick. An unidentified Type 2 Bo-Bo diesel is seen crossing with a down service en route to Hawick. Like many other structures on the southern half of the line the viaduct was short-sightedly demolished in 1985/6, thereby blighting the prospects for the eventual reinstatement of this part of the route. *(Norris Forrest)*

Above. Peak class No. D27 came to grief to the north of Newcastleton whilst working train 1M88, the 10.15 from Edinburgh to St. Pancras 'The Waverley', on 13 June 1964. It is not known whether the photographer was travelling on the train or just happened to be in the right spot at the right moment but he was able to capture this dramatic scene with smoke pouring from the locomotive. Judging by the number of passengers, including some toting cameras and a lady in a headscarf, who can be seen by the lineside in the following image he was probably amongst those unfortunate travellers held up by this conflagration. Today such lineside invasions would no doubt be banned no matter how fascinating the incident. *(W. A. C. Smith)*

Right. Help was summoned and arrived in the very appropriate form of A3 No. 60077 named *The White Knight* riding to the rescue. After the crew had extinguished the blaze, the Pacific deposited the ailing diesel in a siding at Newcastleton station and took the train on tender first to Carlisle. The A3, which had been commandeered from the 11.15 Carlisle – Millerhill Class 5 freight working and had conveniently been in the vicinity at the time of the mishap, eventually worked north again on a military special from Liverpool to Barry Links near Carnoustie in Angus taking over from Jubilee class no. 45627 *Sierra Leone* at Carlisle at 18.25. All in all it was quite a memorable journey for 'The Waverley' passengers that day. *(W. A. C. Smith)*

Left. 1965/66 Timetable.

Carlisle - Hawick - Galashiels - Edinburgh

STEELE ROAD

Above. The positioning of 'The Waverley' headboard varied, sometimes carried above the smokebox number or as here above the buffer beam by A3 No. 60079 *Bayardo* as it passes Steele Road station on 5 March 1960. At this point the train is half way down the 10 miles of falling 1 in 75 grade from Whitrope summit. The photographer recorded in his notes that smoke was coming from one of the carriage wheel-sets indicating that presumably the brakes had been binding for some miles. As well as checking for the presence of rear tail lamps, to confirm the completeness of the train, signalmen were also required to look out for such indications so hopefully this was spotted before any lasting damage was done or a fire ensued. *(Mike Mitchell)*

Right. Steele Road signalbox situated on the down side of the line was a replacement for the original box which had been destroyed by fire in 1914. The replacement box remained operational until January 1965 when it was taken out of use following closure of the goods yard the previous month. Remarkably the station remained staffed for a further two years until March 1967 even though there were only two departures daily in each direction by 1965, all helping no doubt to reinforce the economic case for closure. *(Norris Forrest)*

The Waverley
Restaurant Car Express
LONDON ST. PANCRAS and EDINBURGH WAVERLEY
WEEKDAYS

		am			am
London St. Pancras	dep	9*15	Edinburgh Waverley	dep	10*15
Nottingham Midland	"	11 17	Galashiels	"	11* 5
		pm	Melrose	"	11*11
Chesterfield Midland	"	12 1	St. Boswells	"	11*19
Sheffield Midland	"	12 23	Hawick	"	11*35
Leeds City	"	1 33			pm
Skipton	"	2 8	Newcastleton	"	12 9
Hellifield	"	2 23	Carlisle	"	12 51
Settle	"	2 31	Appleby West	"	1 26
Appleby West	"	3 23	Settle	"	2 12
Carlisle	arr	3 52	Hellifield	arr	2 21
Newcastleton	"	4 34	Skipton	"	2 35
Hawick	"	5 8	Leeds City	"	3 8
St. Boswells	"	5 28	Rotherham Masborough	"	4 18
Melrose	"	5 37	Sheffield Midland	"	4 29
Galashiels	"	5 43	Chesterfield Midland	"	4 54
Edinburgh Waverley	"	6 34	Nottingham Midland	"	5 38
			London St. Pancras	"	7 45

*—Seats may be reserved in advance on payment of a fee of 2s. 0d. per seat.

Left. The Waverley Timetable 1963/64.

Haymarket based A3 No. 60094 *Colorado*, named after the winner of the 1926 2000 Guineas, is seen having just called at Steele Road station, judging by the small number of people visible on the platform, on an unrecorded date in 1960 with an Edinburgh bound service. Steele Road was a very small community with only a few scattered farms in the locality together with two short rows of cottages that post dated the coming of the station in June 1862.
(Neville Stead Collection)

Dismantled signalling equipment lies on the ground to the rear of the up platform in this view of the station yard at Steele Road. Present in the yard is the single decker bus providing a connection to Bellingham. This service, subsidised by the BTC following the withdrawal of train services over the line from Riccarton Junction to Hexham in October 1956, was provided by local bus operator Norman Fox of Falstone. Operating between Bellingham and Kielder it was extended to Deadwater and Steele Road station on Mondays and Saturdays as indicated in the timetable extract, Steele Road being chosen as a railhead because there was no road access to Riccarton Junction.
(Norris Forrest)

Carlisle - Hawick - Galashiels - Edinburgh

This close up of NHN 89 an ex United Automobile Services Bristol L5G single decker with a B35F body was one of the early examples of rural buses converted to one man operation. It waits, probably in vain, for any connecting passengers from the train at Steele Road. This vehicle was acquired by Norman Fox in late 1964 or early 1965 but by 1967 Steele Road was no longer served by road, the bus service by that date terminating at Kielder, thus helping to date this image to the mid 1960s. *(Norris Forrest)*

Table 22 — **RICCARTON and HEXHAM**

Riccarton Junction
Deadwater
Kielder Forest
Lewiefield Halt
Plashetts
Falstone
Thorneyburn
Tarset
Bellingham (N. Tyne)
Reedsmouth
Wark
Barrasford
Chollerton
Humshaugh
Wall
Hexham

> The passenger Train Service between Riccarton and Hexham has been withdrawn. Omnibus Services operated by Norman Fox depart Steele Road on Mondays and Saturdays only at 10 15 am and 5 0 pm serving Deadwater, Kielder Forest, Lewiefield, Plashetts, Falstone, Thorneyburn, Tarset and Bellingham. Charlton Buses (Mid-Tyne Transport) provide connecting services at Bellingham for intermediate places to Hexham.

The aforementioned cottages are evident in this view looking south as an unidentified Brush/Sulzer Type 4 powers northwards. Although appearing to be a rather idyllic setting with smoke from the cottages drifting lazily upwards this could no doubt be a pretty bleak location in the depths of winter when snow was on the ground. Indeed in the severe winter of 1962/3 the snowploughs were out in this area and on one occasion a triple headed plough team consisting of two Black Fives and a 4F 0-6-0 were noted doing battle with the elements between Steele Road and Riccarton. *(Norris Forrest)*

RICCARTON JUNCTION

And so we come to lonely Riccarton Junction, that outpost in the wilds where the Border Counties Railway (BCR) from Hexham connected with the Waverley route. With its smokebox suitably adorned with a wreath, Blaydon depot's K1 class 2-6-0 No. 62022 has arrived at the junction on 13 October 1956 with the final passenger service from Hexham which continued on northwards from Riccarton to Hawick. This was not quite the end of the BCR line however as freight traffic continued until complete closure came on August 22 1958 when goods services were withdrawn between Hexham and Riccarton Junction. *(Neville Stead Collection)*

This view looking south shows the down platform which contained the all important lifeline of a telephone box, no doubt vital for the families living in the environs of the station. Also provided on the station platform was a village shop, open from 9.30am until 2.30pm, and a sub Post Office was installed in the Booking Office in an effort to alleviate the isolation that railway staff and their families must have inevitably felt when based here although at certain times of the year it must have been an idyllic spot. However, the "Scotsman" newspaper once described it as "The Scottish equivalent of a Wild West frontier town: a community created in the middle of nowhere to accommodate the iron horse". For the occasional passenger changing trains here a refreshment room was provided which also did duty as the local pub. The adjacent railway village contained a terrace of houses plus a school and meeting hall.
(Norris Forrest)

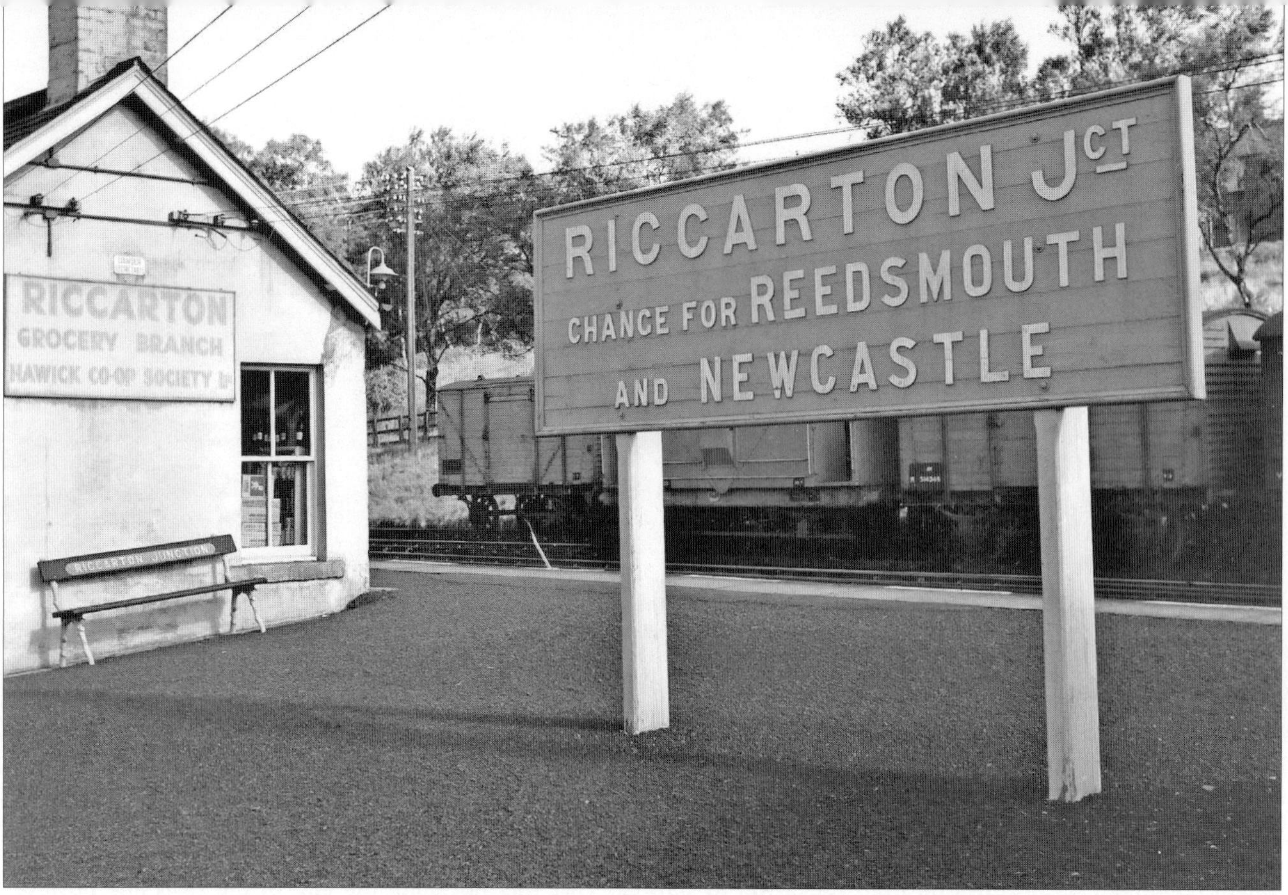
RICCARTON J^CT
CHANGE FOR REEDSMOUTH
AND NEWCASTLE
RICCARTON
GROCERY BRANCH
HAWICK CO-OP SOCIETY Ld
RICCARTON JUNCTION

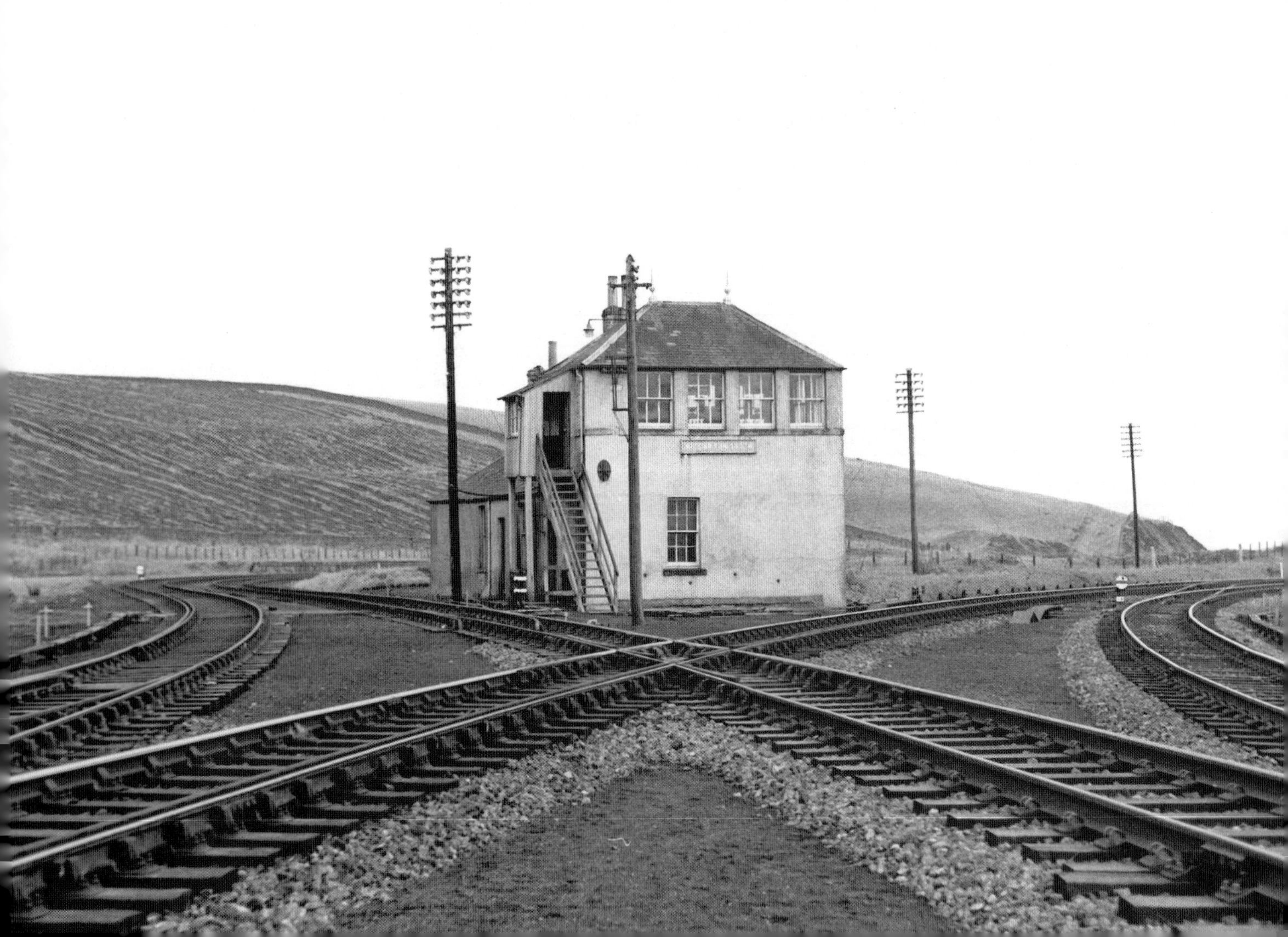

Top left. This close up of the village store reveals that it was a branch of the Hawick Co-op which had a fascinating history. Founded by the Hawick Chartist Association in 1838 the store in the town was opened the following year some five years before the more famous "Rochdale Pioneers" opened a similar venture in England. Merging with the Co-operative Group in 2008 it was, at the time, the oldest continuously trading consumer co-operative in Scotland. Stock for the shop at Riccarton together with the post came in by rail. Also of interest is the large running in board indicating that one could change here for Reedsmouth and Newcastle, via Hexham, until 1956. Note also the seatback adorned with the location necessitating a pretty lengthy sign. *(Henry Priestley)*

Bottom left. Opening in 1881 Riccarton South signalbox controlled the parting of the ways, to the left the single track for Hexham and to the right double track for Carlisle. Although it was to a non standard design it did have the usual sash windows and was constructed in brick although this was rendered no doubt as protection against the sometimes fearsome weather experienced hereabouts. As with Riccarton North box the lever frames were provided by the Railway Signal Company rather than by Stevens the NBR's more normal supplier. Closing with the line on 6th January 1969 this was one of only a pair of boxes that were not subsequently demolished remaining in an increasingly decrepit condition until about 1990 when it is understood it was conveniently made to 'self-destruct' permitting the metal in the lever frame to be recovered for scrap, allegedly by itinerants. *(Norris Forrest)*

Above. A small corrugated iron one road locomotive shed, a sub shed of Hawick, was provided on the south side of the station at Riccarton together with an adjacent coaling stage and in this view D30 Class No. 62440 *Wandering Willie* waits for its next duty. These 4-4-0s were used on Waverley local services and on the Hexham branch, this example being a particular favourite of local enginemen. The shed closed in October 1958 commensurate with the freight closure of the Hexham line. The isolation of the shed was reflected in the fact that it was not included in the traditional enthusiasts' Locoshed Directories. It normally housed two locomotives, one on a 24 hour three shift duty for assisting trains up the bank from Newcastleton to Whitrope and for shunting in the extensive goods yard located here, whilst the other locomotive usually worked an early service to Carlisle and a later goods to Reedsmouth Junction. *(Neville Stead Collection)*

Above. A 55 foot turntable was provided at Riccarton beyond which a PW trolley is parked at the buffer stops. The extensive snow fencing provided here is apparent in this view as is the gradient post indicating a downgrade to the turntable of 1 in 200 with an upgrade of 1 in 100 leading away from it. The turntable was the largest on the line between Carlisle and Edinburgh and locomotives that were unable to turn on Hawick's smaller facility made the journey to Riccarton to accomplish this. *(Norris Forrest)*

Top right. Lineside snow fencing is again in evidence as B1 Class No. 61397 gets away northbound with a two coach local service for Hawick. Built by the North British Locomotive Co. in Glasgow it was allocated to St. Margaret's depot from new in 1952 and remained at the Edinburgh shed until withdrawal in June 1965 after just 13 years service. *(Neville Stead Collection)*

Bottom right. Being so isolated Riccarton was an ideal place to store unwanted stock out of the glare of public gaze and sidings here were often used for this purpose. In this view from the early 1950s C15 class 4-4-2Ts Nos. 67472 and 67477 are in store on part of the Hexham branch, their working days almost certainly over. Designed by Reid and introduced in 1911, the last two survivors of this class of 30, Nos. 67460 and 67474, managed to cling on until April 1960. *(Neville Stead Collection)*

Carlisle - Hawick - Galashiels - Edinburgh

WAVERLEY

40

Top left. A southbound parcels working hauled by a Peak diesel passes a northbound freight hauled by V2 class No. 60970 putting on a volcanic display in the vicinity of the former Riccarton North signalbox which had opened in 1881, only to be burnt down and subsequently rebuilt in the 1940s. Unlike its southern neighbour the north box was not rendered being in plain brick, no doubt due to the protection rendered by other buildings and the adjacent hillside. It finally closed in April 1959. This viewpoint gives a good impression of the terrace of houses above which, on Saughtree Fell, the Forestry Commission has been busy ploughing for future afforestation. *(Norris Forrest)*

Bottom left. Not the most ideal conditions for a photographic stop perhaps but on 14 April 1963 A3 Pacific No. 60041 *Salmon Trout* had charge of a portion of the 'Scottish Rambler No. 2' railtour which has come to a stand at Riccarton Junction's southbound platform at approx. 7.20pm. This was a multi engined tour involving no less than fourteen locomotives spread over four days. On this, the third day of the tour, No. 60041 had taken over from No. 61324 at Hawick, the B1 having handled the leg from Coldstream to Hawick taking in Kelso, Jedburgh and St. Boswells on the way, and proceeded south to Carlisle. *(W. A. C. Smith)*

Above. Our final view of Riccarton dates from 18 August 1967 by which time DMUs, in the shape here of a Derby Lightweight unit, had taken over most of the local workings on the line. The driver of the service seen at the northbound platform awaiting departure for Hawick leans out of his cab window no doubt amazed at the prospect of a photographer at this remote spot. Its remoteness had been somewhat alleviated by the laying down of a forestry road in 1963 which led to local inhabitants having to lock their doors at night for the first time, making road connection to the wider world something of a mixed blessing. *(Henry Priestley)*

WHITROPE

Above. Class V2 No. 60922 slogs up to Whitrope summit on 4 June 1960 with a down passenger working consisting of six coaches. The train will shortly enter the 1208 yard long Whitrope tunnel, the fourth longest in Scotland. Although it is a Category B listed structure it has suffered from a couple of partial collapses in the 21st. century. In March 2002 there was a partial collapse of the tunnel roof at the south portal followed by a major collapse in March 2021. Although undoubtedly repairable there is currently no source of funding for this, which has resulted in the sealing of the tunnel for safety reasons. In the foreground are yet more snow fences indicating the exposed nature of the line at this point. *(Neville Stead Collection)*

Right. A signalbox was provided at Whitrope Siding which was used to stable banking locomotives awaiting return to either Hawick or Newcastleton after having assisted trains up the gradients to the 1006 foot summit of the line. The siding closed in February 1964 whilst the box remained open until November 1967. There was an unadvertised halt for railway staff that lived nearby, and for use by the public until the February 1964 closure, although in the absence of any platform access to trains was by means of a step ladder. *(Norris Forrest)*

WAVERLEY

SHANKEND

Top left. Also on 4 June 1960 the photographer has captured Class K3/2 No. 61988 erupting from the southern portal of Whitrope tunnel with an up freight service. These 2-6-0s were a common sight on freight workings along with V2s and A3s and this example which dated from 1935 was allocated to Sheffield Darnall shed at this date but would only last in traffic until the end of the following year before withdrawal. The substantial brick supporting walls of the cutting side are apparent in this view. *(Neville Stead Collection)*

Bottom left. Producing a spectacular visual display as well no doubt an aural one, Heaton based Class V2 No. 60835 previously bearing the lengthy nameplate *The Green Howard, Alexandra, Princess of Wales's Own Yorkshire Regiment* leaves Whitrope tunnel and enters the deep cutting with a southbound freight, assisted in the rear by a banking locomotive which will drop off and retire to the siding to await a return path. These Gresley 2-6-2s gave sterling service to the Waverley route for more than 30 years and during the 1960s, following the opening of the marshalling yards at Carlisle Kingmoor and Edinburgh Millerhill, which were to prove very short lived, they played a large part in handling the hourly freight services which often operated over the full 24 hour period. Schedules were less than demanding however with heavy freights taking anything from four to seven hours to travel the entire route. One of the few dwellings in the vicinity can be seen on the hillside on the right. *(Norris Forrest)*

Above. The hillside location of Shankend signalbox and station is apparent in this view of a lengthy southbound freight service toiling up the grade to Whitrope summit. On the adjacent slopes cows graze unfazed by the sight and sound of the passing train. The small hamlet situated nearby can have contributed little to passenger revenue and such isolated stations as these will surely not be resurrected should the line be reopened south of Hawick in the future. *(Norris Forrest)*

Above. This 1953 image reveals the attractive station gardens whose care must have been challenging in such a windswept location but doubtless station staff had plenty of time between services to tend them. However, staff would be withdrawn from here in July 1961 with the station being closed to goods traffic at the end of 1964. By 1965 just two services daily were provided in each direction to Carlisle and Edinburgh supplemented by a third (SX) service to Hawick and a late departure (SO) from Edinburgh to Newcastleton depositing passengers at Shankend at the rather ungodly hour of 49 minutes past midnight on Sunday morning. *(Neville Stead Collection)*

Top right. This view looking south reveals that a footbridge had subsequently been provided, in the early 1960s, behind which is a grounded coach body mounted on a brick base. In the distance beyond the tall NBR lattice signal can be seen the signalbox whilst the green fingers of the staff seemingly extended to this end of the station also where another attractive floral display enhances the scene. *(Henry Priestley)*

Bottom right. Passing the aforementioned tall signal comes Class A3 No. 60043 *Brown Jack* with a down express in an undated image. This Gresley Pacific spent much of its working life based in Edinburgh at Haymarket and latterly St. Margaret's sheds before withdrawal in May 1964. There was a small goods yard off to the left consisting of three loop sidings, accessible from both ends, together with a cattle dock which handled freight traffic for the surrounding area of hill farms and isolated dwellings. The goods yard also housed banking locomotives on occasions to assist trains on the climb up to Whitrope. *(Neville Stead Collection)*

Carlisle - Hawick - Galashiels - Edinburgh

47

For the last few years of the local passenger service DMUs played a significant role and here a Derby Lightweight 2 car unit is seen passing the signalbox at Shankend with a service to Hawick under the watchful gaze of the signalman. The box dated from 1916 replacing the original 1888 structure and of note is the small shunting signal on the left permitting egress from the goods yard sidings. The signalman can hardly have imagined that half a century later in 2013 his box, now converted to a holiday home, would have sold for a reported price of £90,000. *(Henry Priestley)*

Before its conversion however the box took on an unusual role for a number of years as a store for hay bales and is seen here in this capacity in the late 1980s. Much of the ballast remains in situ although now covered in vegetation some 20 years having passed since closure in 1969. *(Jeffery Grayer)*

Carlisle - Hawick - Galashiels - Edinburgh

One of the notable engineering features of the line was undoubtedly the 15 arch Shankend Viaduct seen here being crossed by a southbound freight service hauled by a Black Five laying down a spectacular exhaust. The structure is nearly 600 yards long, 60 feet tall at its highest point and crosses the Langdale Burn. It was extensively restored in the early years of this century by BRB (Residuary) Ltd. Of interest in the train's consist is the first vehicle behind the tender which appears to be a small diesel shunter. These were sometimes seen in later years travelling in the opposite direction en route to Scottish scrapyards such as Campbells of Airdrie as several of these short lived Hunslet Class 05 0-6-0s are known to have made their last journeys over the Waverley line in the late 1960s. *(Norris Forrest)*

WAVERLEY

STOBS

HAWICK

Top left. Stobs was another of those isolated wayside stations and is seen here looking south in 1955. Although there was little habitation in the vicinity, there was a military camp set up by the War Office in 1903 close by which boasted its own narrow gauge tramway. After performing a number of functions including acting as an internment centre for civilians during WW1, a resettlement centre for Polish soldiers of WW2 and after 1947 a summer camp, it was sold off in 1959. Stobs signalbox seen here on the down platform closed in December 1962 some 18 months after the goods yard had closed and the station became unstaffed. *(Neville Stead Collection)*

Bottom left. Time is nearly up for the Waverley route as the presence of camera toting enthusiasts on the platform and on the embankment together with the number of heads poking out of the carriage windows might signify for this view looking north dates from November 1968, a few weeks before closure. An unidentified BRCW Type 2 diesel leaves Stobs for Hawick and Edinburgh and will shortly cross Barns Viaduct located immediately to the north of the station. *(W. A. C. Smith)*

Above. A4 Pacific No. 60004 *William Whitelaw* heads the 2.36pm Edinburgh Waverley – Carlisle on 26 September 1960 away from the southern outskirts of Hawick. Allocated to Haymarket depot at this date it would go on to be part of that 'Indian Summer' of the A4s when it was transferred to Aberdeen Ferryhill in 1962 to work the 3 hour expresses to Glasgow before withdrawal in July 1966. Down in the valley can be seen the roof of the famous Lyle & Scott clothes and fabric manufactory still in operation today. *(W. A. C. Smith)*

And so we come to the first town of any size encountered since leaving Carlisle – Hawick located on the River Teviot 45½ miles from the Cumbrian city. This view of the station approach includes the magnificent departure board with its enamels indicating times of trains to such destinations as Galashiels, Alloa, Falkirk, Montrose, Dundee, Glasgow, Inverness, Kirkcaldy and Edinburgh. A split screen Morris Oxford on the left, a trolley piled high with parcels outside the station entrance and a couple of bonneted ladies deep in discussion on the right complete this charming period scene. (*Arthur Mace*)

Crossing the River Teviot on a substantial viaduct, with North Road bridge in the background, is A1 Pacific No. 60161 appropriately named *North British* with the 12 noon Waverley – Carlisle service which has just departed from Hawick station on 26 September 1960. Stopping only at Riccarton, Steele Road and Newcastleton it would reach Carlisle at 2.53pm. Advertisements of note on the hoarding include Usher's Export pale ale, brewed at Park Brewery in Edinburgh, and Player's Bachelor cigarettes which in 1956 for example cost 3/3d (16p) for 20! *(W. A. C. Smith)*

Worsdell D20 Class 4-4-0 No. 62387 waits at Hawick in this undated view. Built in 1907 it would last in traffic until withdrawal in September 1957, the year which saw the withdrawal of the few remaining members of the class. It is carrying shedcode 52B Heaton which helps to date this view to the period between June 1953 and July 1955. The section of wooden planking, installed to reduce weight on the viaduct, indicates those sections of the platforms which were situated on that structure extending over the river. *(Neville Stead Collection)*

Above. Standing in a similar position but photographed this time from the up platform is Glen Class D34 No. 62490 *Glen Fintaig* again in an undated view with a stopping service to Carlisle. The eponymous glen is situated in Inverness-shire north of Spean Bridge. Carrying its class number on the buffer beam together with details of its home depot St. Margaret's helps to date this view to 1958/59 shortly before withdrawal from 64A which came in December 1959. Hawick's substantial station buildings, a covered lattice structure for passengers and goods which replaced an earlier footbridge and the top of the tall signalbox are evident in this image. *(Neville Stead Collection)*

Bottom left. The rear of the tall South signalbox is seen here, its height having been raised in 1913 to afford the signalman a better view over the parcels hoist used by Post Office and station staff to transfer trolleys from one platform to another. There was a short bay platform to the left which acted as a goods dock but was also used for some northbound passenger services originating at Hawick. North signalbox which was closed in 1965 controlled the northern approaches to Hawick whilst the South box controlled access to the locomotive shed and goods yard. It remained open until the end of passenger services, subsequently being demolished in 1972 with demolition of the station and goods yard following in 1975. The site is now occupied by the Teviot Leisure Centre although the trackbed north of the station is reasonably free of later developments, so connection to the Borders Railway at Tweedbank is certainly feasible given the political will and of course the finance. *(Henry Priestley)*

Above. Class J36 to a Holmes North British design dating from 1888 No. 65232 is seen at the north end of Hawick station in this mid 1950s view. It was transferred from Hawick shed at the end of 1956 to Polmadie and thence to Kipps depot ending its days there in October 1961 having given 70 years of useful service. Several of the members of the class were given names associated with the First World War but not this example. One, No. 65243 *Maude*, has been preserved and can be found on the Bo'ness & Kinneil Railway in West Lothian. *(Neville Stead Collection)*

WAVERLEY

Two small boys, who bear a striking resemblance to the pair seen on the front cover, have apparently migrated to the up platform to watch the arrival of Class V2 No. 60970 with a lengthy down goods service. The elegant lines of Gresley's graceful 2-6-2s are shown to advantage in this view which also features again the tall South signalbox and one of the two substantial timber faced towers which contained goods lifts and which were connected by a footbridge. The two road motive power depot, coded 64G and seen on the right, had an allocation of 11 locomotives in January 1961. (*Sandy Murdoch*)

A4 Pacific No. 60007 *Sir Nigel Gresley* paid a visit to the line in 1964 when it handled a leg of the RCTS 'Scottish Lowlander' tour of 26 September. Originating from Crewe with No. 46256 *Sir William A Stanier F.R.S.* at the helm it changed locomotives at Carlisle where the A4 took over for a run over the Waverley route to Niddrie West Junction where fellow A4 No. 60009 *Union of South Africa* returned the tour to Carlisle via Kilmarnock and Dumfries from where the Coronation proceeded to Crewe. With 450 tons in tow the performance of the A4 over the Waverley route was described as outstanding, being an all time record for a run over the line with that trailing load. Seen here during the 10 minute stop at Hawick several of the tour participants can be seen obtaining their photographic souvenirs of the day. *(Neville Stead Collection)*

Two years later and another Pacific graces Waverley metals with a railtour heading south this time hauled by Class A2 No.60532 *Blue Peter* with the BR (Scottish Region) tour of 8 October 1966. Starting from Edinburgh this tour headed over the Waverley route to Carlisle returning via Beattock. This was to be its final railtour as it was withdrawn from Aberdeen Ferryhill shed at the end of the year. Going into storage it was purchased for preservation, being the sole example of its class. *(W. A. C. Smith)*

WAVERLEY

HAWICK SHED

Top left. Earlier in 1966, on 25 June to be exact, the most famous locomotive of all No. 4472 *Flying Scotsman* visited the Waverley line with 'The Aberdonian' railtour which it hauled from Hellifield to Edinburgh. This tour operated over three days from 24-26 June featuring a diverse mix of motive power including a Merchant Navy, an A4, a V2 and a 9F. At Edinburgh the A3 handed over to a pair of J37s for the onward leg to Anstruther, Leven and Thornton Junction. No. 4472 adorned with a suitable headboard is seen here at Hawick just before midday prior to departure for Edinburgh. *(Trevor Davis)*

Bottom left. Our final view of Hawick station is a much sadder occasion as this is 23rd. November 1968 just six weeks before closure. On this day the Border Railway Society organised a 'Farewell to the Waverley Route' railtour hauled by BRCW Type 2 No. D5311. The headboard would be used again for on 4 January 1969, the final Saturday of passenger services, D5311 worked up from Carlisle in the morning on the 9.20am scheduled service departure sporting the headboard returning later that afternoon. The following day Deltic D9007 *Pinza* would operate the formal 'Farewell to the Waverley' RCTS railtour run from Leeds over the Settle & Carlisle route. *(W. A. C. Smith)*

Above. We now turn our attention to Hawick shed and yard where a number of veterans could still be seen in the early 1960s such as this Glen D34 Class No.62484 *Glen Lyon* photographed by the water tower in 1962. Sadly this locomotive had been withdrawn in November the previous year and was out of use having spent its last year in service based at Hawick. It was not subsequently scrapped until May 1963 when the firm of Arnott Young based in Old Kilpatrick got to work on this 4-4-0 some 44 years after it had been constructed back in 1919. *(Neville Stead Collection)*

Above. Another veteran in sparkling external condition was Class D30 No. 62425 *Ellangowan* seen in the yard in this view which, although unrecorded by the photographer, probably dates from the late 1950s as the locomotive was withdrawn in July 1958. The locomotive had been a long time Hawick resident and was very popular with crews hence its immaculate appearance. *(Neville Stead Collection)*

Top right. More modern motive power also utilised Hawick shed's facilities as evidenced by this Carlisle Canal based Class D49 No. 62732 *Dumfries-shire* seen here in 1952. One of the Shire/Hunt class Gresley designed 4-4-0s it was fitted with a Great Central tender and was one of only four members of the class whose nameplate originally contained a hyphen between the county name and the word 'shire' the others being Kinross, Peebles and Inverness. The original nameplates fitted to No. 62713 *Aberdeenshire* had the name as a single word without a hyphen but a hyphenated plate was sold on the railwayana market in recent years. This was presumably a replacement fitted when the locomotive visited Cowlairs Works and the original plates were found to be damaged, resulting in replacement plates being cast with a hyphen, no doubt Cowlairs Works considering this to be more in keeping with the hyphenated plates carried by other members of the class. The D49s were not very popular with Scottish Region footplate crews, however, who often complained about their poor riding quality and draughty cabs. *Dumfries-shire* was condemned in November 1958. *(Neville Stead Collection)*

Bottom right. A couple of the short lived Standard Class 2 2-6-0s were allocated to Hawick from new in 1955 including No. 78047 photographed from the down platform at Hawick station. They were joined by further members of the class in 1959 and 1960 with the final example, No. 78049, leaving Hawick in January 1966 upon closure of the shed. Afterwards the shed, which had opened in 1849 and been re-roofed in 1955, was demolished but remained as a signing on point until closure of the line in 1969. *(Sandy Murdoch)*

Carlisle - Hawick - Galashiels - Edinburgh

WAVERLEY

Top left. A pair of Hawick stalwarts of the D30 'Scott' Class Nos. 62428 *The Talisman* and 62423 *Dugald Dalgetty* was captured on shed in this undated view. Although the first three of the class were withdrawn between 1945 and 1951 and despite the introduction of large numbers of the more powerful B1 class, withdrawal of the remainder did not restart until 1957. The end of 1958 saw the withdrawal of No. 62428 whilst No. 62423 had met its fate a year earlier in December 1957 and although the last two class members made it into 1960, sadly none were preserved.
(Neville Stead Collection)

Bottom left. Allocated to Hawick in July 1961 Standard tank No. 80113, seen here in front of the substantial water tower situated at the south end of the locomotive shed, was a regular performer on the 6.30am Hawick to Carlisle passenger working and the 18:13 (SX) return during 1964. By late 1965 Hawick's allocation had been reduced to just two locomotives Nos. 80113 and 78049, the latter being generally used on station pilot duties when the usual Clayton diesel was unavailable as it was being used to cover for a failed machine on the mainline. The Class 4 tank is buffered up to an old coach that resided in the yard for many years. *(Neville Stead Collection)*

Above. Our final view of Hawick shed, taken on 9 June 1960, reveals Class C16 4-4-2T No. 67489 and Class J36 0-6-0 No. 65316 and a small section of that ancient coach mentioned previously. A goodly supply of snowploughs lies on the ground awaiting the winter's inclement weather which often interrupted operations on sections of the exposed Waverley route. The shed was a two road affair with just the one exit at the south end seen here. Inside a couple of small boys are being given a guided tour hopefully avoiding the inspection pit hazard. The C16 would be withdrawn in February the following year whilst the J36 would soldier on until the end of 1962. *(Henry Priestley)*

HASSENDEAN

V2 No. 60808 rumbles through Hassendean with a down freight service in 1960 passing the signalbox, the occupant of which has already returned the tall signal behind the locomotive to danger. The height of this signal was undoubtedly dictated by the presence of the overbridge seen in the distance. Track through the station was on the level but climbs of 1 in 150 and 1 in 175 would follow up to the summit near Standhill. The signalbox closed in February 1965 whilst No. 60808 was scrapped at Darlington Works in October 1964. *(Neville Stead Collection)*

This view looking north and taken sometime during 1967/8 postdates the station being downgraded to unstaffed halt status, which occurred in March 1967, as witnessed by some of the vegetation which is in need of a trim. Whilst the goods yard sidings had closed back in December 1964 the numerous oil lamps mounted on the platform on short posts remained an attractive feature. Today the station, which still retains its footbridge, is in private ownership providing luxury self catering cottage accommodation. *(Henry Priestley)*

Although Hassendean station first appeared in public timetables in 1850 a signalbox was not provided here until 1882 when it was installed to control access to the small goods yard. Sadly it was demolished shortly after it closed in 1965 as part of the retrenchment policy which saw small goods yards and several signalboxes closed along the length of the line. *(Norris Forrest)*

Above. A3 Pacific No. 60079 *Bayardo* was photographed near Hassendean in 1960 with an unidentified passenger working. One of four A3s based at Carlisle Canal shed at this time No. 60079 would be withdrawn in September the following year and was noted in the scrapline at Doncaster Works together with classmate 60035 *Windsor Lad* the following month. Note the immaculately maintained lineside vegetation and the manicured ballast of the "permanent way" which in nine years' time would prove to be anything but. *(Neville Stead Collection)*

Top right. Belses station, opening as New Belses after an adjacent farm in 1849 but renamed plain Belses in 1862 is seen here looking south in 1955. Oil lamps on short posts are again in evidence and were a feature of many of the smaller stations on the line. The goods yard remained open until December 1964 with the station losing its staff in March 1967 when most of the stations not already unstaffed were made into unstaffed halts leaving only Galashiels, Melrose, St. Boswells, Hawick, Newcastleton and Penton with staff. *(Neville Stead Collection)*

Bottom right. Class B1 No. 61396 gets under way after its stop at Belses with the 8.05am service from Hawick to Edinburgh on 20 February 1965. After stopping at all stations to Heriot it would then omit three stations calling finally at Eskbank & Dalkeith before terminating at Waverley station at 9.43am. This was no doubt a popular service with passengers from the Border towns wishing to spend a day's shopping in the capital there being convenient return services from Edinburgh at 4.10pm and 5.54 pm. *(Norris Forrest)*

Carlisle - Hawick - Galashiels - Edinburgh

BELSES

Above. As Belses itself consisted of just a couple of farms and a few cottages the large running in board drew attention to the fact that this was the station for Ancrum and Lilliesleaf. What it failed to mention however was the fact that neither place was what you might call significant and that Ancrum, current population 300, was 2½ miles away and Lilliesleaf, of a similar size, was even further at some 3½ miles, the latter village being equidistant from the alternative station at Hassendean. *(Norris Forrest)*

Left. In common with many boxes in Scotland the signalbox name was not on the front of the box but mounted on the side wall. As can be seen the box, like many on the line, was constructed entirely of brick with no wooden superstructure – no doubt a reflection of the inclemency of the prevailing weather conditions. Opening in 1882 it closed at the end of July 1966. Of note is the attractive finial on the roof. *(Norris Forrest)*

ST. BOSWELLS

In this northward view of St. Boswells the substantial covered footbridge is evident whilst the running in board indicates that this was once the junction for trains to Kelso and Jedburgh, the latter destination now replaced by a bus service. The branch to Kelso, Tweedmouth and Berwick-upon-Tweed was to close in 1964 whilst that to Jedburgh had been closed due to flood damage back in 1948. At one time one could also change here for Duns and for Reston situated on the ECML but this route had closed as far as Duns due to damage caused by the severe floods of August 1948 (details of this event can be found in issue 1 of the Transport Treasury publication 'Railway Times'). St. Boswells was also the railhead for visitors to nearby Dryburgh Abbey, last resting place of three famous Scotsmen – Sir Walter Scott, Earl Haig and the novelist John Buchan. *(Henry Priestley)*

On 12 June 1952 St. Margaret's allocated No. Class K3/2 No. 61857, one of the class fitted with a Great Northern tender, rattles through St. Boswells with an up goods service passing the shed yard on the right where an unidentified Class J39 is simmering quietly. This 0-6-0 had probably worked in on the previous night's goods service from Carlisle and it would then handle a freight turn to Kelso and Jedburgh before returning to Carlisle with a mid evening goods working. *(W. A. C. Smith)*

WAVERLEY

Twelve years and one day later to be exact, on 13 June 1964, Standard Class 4 No. 76050 was noted on a down passenger working attracting some custom at St. Boswells. This Mogul had arrived at Hawick depot in November the previous year where it would remain until withdrawal in September 1965. The projecting overhang of the signalbox is seemingly providing some shelter for at least one passenger on what looks to be a rather damp day. The large running in board on the left appears to have lost its letters leaving behind just the ghostly impression of past rail served destinations. *(W. A. C. Smith)*

This close up of St. Boswells South signalbox reveals the extent of the overhang and shows the convenient placing of a bench adorned with the station name on the back. It also features a rather fine lamp illuminating that area of the down platform. On its right hand side the box overlooked the bay platform formerly used by Duns branch trains. *(Norris Forrest)*

The Kelso branch train headed by D30 'Scott' class 4-4-0 No. 62440 *Wandering Willie* in the bay platform is readied for departure in this 5 June 1954 view of the southern end of the station. On the right is the small locomotive shed, a sub shed of Hawick, which is currently housing Class J39 No. 64733. Trains on the Kelso line were never frequent and even back in 1895 there were only 5 departures (SuX) each way which had reduced to 4 by the date of this view. But a year later drastic economies were made and only two departures each way were provided, with the train crew of three often outnumbering the passengers in the single coach provided until the inevitable closure of the branch in June 1964. *(Neville Stead Collection)*

WAVERLEY

Top left. A trio of old timers is seen outside St. Boswells shed in 1956. They are from left to right Class J35 0-6-0 No. 64499 dating from 1909, Class D30 Nos. 62440 *Wandering Willie* and 62423 *Dugald Dalgetty* dating from 1920 and 1914 respectively. It is appropriate that members of this 'Scott' class of 4-4-0 should find work on the Waverley route, often on Hawick to Carlisle locals, named as they were after characters from, or titles of, the Waverley novels. Wandering Willie's tale features in 'Redgauntlet' whilst Dugald Dalgetty can be found in 'A Legend of Montrose'. *(Neville Stead Collection)*

Bottom left. The crew of Class D49 No. 62734 *Cumberland* exchange a few words outside St. Boswells shed in 1953. This 'Shire' 4-4-0 based at Carlisle Canal depot dated from 1929 and was provided with Walschaerts valve gear and derived motion and was one of eight members of the class fitted with a NE tender. It remained at Carlisle until withdrawal in March 1961. *(Neville Stead Collection)*

Above. On this rather damp day in 1956 the Kelso branch was graced with more modern motive power in the shape of Gresley V1 Class 2-6-2T No. 67659 constructed in 1936. Based at St. Margaret's shed this was one of 43 members of the V1 class introduced in 1930 with a further 49 examples of the similar V3 class built, or rebuilt from existing V1s, with a higher boiler pressure from 1939 onwards. It lasted in traffic until February 1962. Hawick shed received one example of each type back in the late 1950s, although whilst allocated to Hawick they were actually sub shedded at St. Boswells to operate the Berwick-upon-Tweed via Duns service. An intending passenger seems to have had a change of heart as he leaves the carriage, whilst at the end of the bay platform a pram awaits loading into the guard's van. *(Neville Stead Collection)*

Above. "Bulled up" for railtour duty 'Glen' D34 Class No. 62471 *Glen Falloch* makes a fine sight at St. Boswells with the Branch Line Society (BLS) 'Scott Country' special of 4 April 1959. This tour had commenced at Galashiels and visited Selkirk before returning to Galashiels and thence to St. Boswells. From here it went to Greenlaw, Roxburgh and Jedburgh before returning to St. Boswells and Galashiels. Tour participants were no doubt taking the opportunity of the stop to photograph the locomotive and to take a quick tour around the adjacent shed. The 4-4-0 would be withdrawn from St. Margaret's shed in March 1960 having given 46 years service. *(W. A. C. Smith)*

Top right. The closure of the shed in November 1959 was followed on 15 June 1964 by the withdrawal of the Kelso branch passenger service. Standard Class 4 No. 76050 was drafted in to handle the final departure for Kelso and Berwick at 4.02pm which was strengthened to 3 coaches rather than the normal solitary vehicle. However this proved unnecessary as the maximum complement was recorded as being 8 passengers of which 3 were enthusiasts and 1 a railwayman. The shed entrance was bricked up and here displays a 'Bibby' label indicating its new role as a storage facility for the agricultural merchants. In later years the yard became an oil storage depot and the shed was subsequently used by a van hire company. Remarkably it remains in situ today. *(Norris Forrest)*

Bottom right. Our final view at St. Boswells was taken during the diesel era and shows Type 2 Bo-Bo, later Class 26, No. D5313 arriving with a service for Carlisle. Constructed by the Birmingham RC&W Co. in 1959, it was based at Haymarket depot from 1960 until transfer to Inverness in November 1975 from where it was withdrawn in March 1986, being one of many of the class scrapped at Vic Berry's yard in Leicester. (For more details of this yard see the author's volume 'Colin Garratt's Scrapbook' published by Transport Treasury). *(Norris Forrest)*

Carlisle - Hawick - Galashiels - Edinburgh

MELROSE

Top left. This post-closure view of Melrose looking south reveals the attractive main station building which was described upon its opening as the 'handsomest provincial station in Scotland'. It was built in a Jacobean style with Dutch gables, transom windows with stone mullions, tall ornate octagonal chimneys and a portico entrance seemingly far in excess of the needs of the limited local population. But the town was developing as a tourist destination at the time of the railway's arrival in 1849. The platform canopy was of an unusual design sloping upwards at some 30 degrees and supported by cast iron columns with lotus capitals. After a period of dilapidation after closure it was eventually awarded the Scottish equivalent of a Grade 1 listing.

Bottom left. Precursor to 'The Waverley' was 'The Thames – Forth Express' which was introduced by the LMS in September 1927 concurrently with 'The Thames-Clyde Express'. Operating between London St. Pancras and Edinburgh Waverley the train took the Midland route over the Settle & Carlisle line and then the ex North British Railway Waverley route to Edinburgh. The title was withdrawn in September 1939 for the duration of the Second World War but the name was never resurrected in peacetime. The service was however re-introduced by BR in 1957 taking the title 'The Waverley'. The pre war version of the express is seen here at Melrose during the 1930s hauled by NBR Reid Class H Atlantic No. 9879 named *Abbotsford*. At the time of their introduction in 1906 they were reputed to be the heaviest, longest, and most powerful locomotives yet employed by the NBR. Giving long service, all had been withdrawn by 1937 although one was put aside for preservation but not until it had been partially scrapped. It was painstakingly rebuilt from the parts which were still in existence but with the outbreak of WW2 the drive was on for metal for aircraft production so the locomotive, No. 9875 *Midlothian*, was scrapped for a second time.
(Dr. Ian C. Allen)

Above. In this 1954 view Scott Class D30 No. 62422 *Caleb Balderstone* rolls into Melrose with a lengthy train which may well be a special working or an excursion as it is carrying express passenger headlamps not often worn by this class of locomotive which were generally utilised for secondary duties. Based at Hawick in the 1950s this 4-4-0 was withdrawn in December 1958. A representation of this Scott character can be found on the north facade of the Scott Monument in Princes Street Edinburgh. Described in 'The Bride of Lammermoor' as an old man with 'thin grey hairs, bald forehead, and sharp high features', he is recorded as having developed the ability to offer hospitality verbally whilst withholding it in practice into a fine art. Possibly a precursor to the stereotypical Scotsman characterised as being 'careful with money'? *(Neville Stead Collection)*

Taken from a similar vantage point this undated view shows Haymarket based Class A2/1 No. 60510 *Robert the Bruce* running into Melrose with an Edinburgh service. One of the sub class of A2s numbering just four examples No. 60510 was a development of the original Thompson Pacifics incorporating a Class V2 boiler. Three were withdrawn in 1960 with the final example going the following year. *(Neville Stead Collection)*

Looking from the up platform it can be seen that the unusual design of canopy was replicated on the down platform behind which was located the goods shed. Freight traffic finished here in May 1964. The Dutch gables of the main station building are evident on the right and the gas globes make an attractive feature illuminating the platforms at night. Today traffic thunders past here on the Melrose by-pass opened by the Chancellor of the Exchequer George Osborne in 2010 at a cost of £52m and built on the site of the former down platform and goods yard. Whilst a great boon to the local residents spared the sight and sound of traffic through the historic town centre it is undoubtedly a great hindrance to any further expansion of the Borders Railway from nearby Tweedbank. *(Neville Stead Collection)*

Carlisle Canal based Class A3 No. 60093 *Coronach* enters Melrose with the 2.36pm Edinburgh to Carlisle service on 25 September 1961. If running to time arrival here was at 3.40pm with Carlisle reached at 5.45pm. The station was well situated for the town being only a short distance from the market place and these days the main building acts as an Italian restaurant. *(W. A. C. Smith)*

Above. On the same date Britannia Class Pacific No. 70018 *Flying Dutchman* was in charge of the 1.28pm service from Carlisle to Edinburgh which would subsequently call only at Galashiels. It is seen here departing Melrose passing the signalbox situated at the end of the down platform with access to the goods yard being visible on the right. A recent transferee to Carlisle Canal shed No. 70018 would ultimately return to Carlisle, although this time to Kingmoor shed after several intervening transfers, from where it would be withdrawn in December 1966. It would make one final journey to Scotland, this time for scrapping at the premises of Motherwell Machinery & Scrap at Wishaw. *(W. A. C. Smith)*

Opposite. A trio of images showing the decline and partial rebirth of Melrose station. The first image dates from the 1980s when vegetation was threatening to swamp the site and the down platform is still in situ prior to the construction of the by-pass. The other two images taken in 2021 show the present well cared for appearance of the station with the surviving platform fitted with seats and replica signage. Hopefully with some adjustments to the by-pass there may just be space to accommodate a new single track should an extension of the Borders Railway receive the go ahead in future years. *(Jeffery Grayer)*

GALASHIELS

82

Top left. St. Margaret's based Class V2 No. 60953 leaves Galashiels in this undated view passing under a bridge with a service for Edinburgh and with a timber yard on the right. This bridge, which still stands and now spans the restored railway, carries the A7 and is known as High Buckholmside but of the former station there is now no trace, a new single platform having been constructed to serve the Borders Railway. *(Neville Stead Collection)*

Bottom left. Running into Galashiels station with a local service for Hawick is Class D30/2 No. 62437 *Adam Woodcock* sporting a Haymarket allocation on its buffer beam. Classified 3P, these 4-4-0s were a common sight over the Waverley route during the 1950s and this undated view was taken prior to transfer to Edinburgh's St. Margaret's shed which occurred in September 1955. This locomotive was withdrawn in June 1958 by which time there were just 13 examples of the class left in service with the last two going in June 1960. Incidentally Adam Woodcock was a character in one of Scott's lesser known novels 'The Abbot' written in 1820. *(Neville Stead Collection)*

Above. This mid 1960s view looking south shows the extent of the former station with the platform loop to the left which was formerly used by trains operating to Selkirk and via the Peebles loop until withdrawal of the service to the former in September 1951 and to the latter in February 1962. The tracks here were crossed not only by the footbridge but also by Station Brae road bridge seen behind which replaced an earlier level crossing. This bridge was replaced in 2007 to accommodate the new railway. The station at one time had an overall roof but this was removed in the 1930s and replaced by canopies braced with the steel girders seen here spanning the tracks behind the footbridge. *(Norris Forrest)*

Running into Galashiels with a local service from the north is Class V3 2-6-2T No. 67670 in this undated view. This was one of the class rebuilt, in June 1956, from Gresley's original V1 to operate at a higher boiler pressure. It was withdrawn from St. Margaret's depot in August 1961 and note that the station's running in board is still proclaiming this as the junction for Innerleithen and Peebles, which would tie in with a potential date for this image between 1956 and 1961. *(Neville Stead Collection)*

This view of the south end of Galashiels station dates from 14 May 1965. The signal post on the left contains three elevated ground disc signals. This angle gives a better view of the girders spanning the tracks and of the road bridge seen climbing the hill to the right. The substantial station building, dating from 1849 and serving the community for 120 years until closure in 1969 and subsequent demolition, is also evident. *(James L. Stevenson)*

In this undated view A3 Pacific No. 60057 *Ormonde* departs from Galashiels with a down service on a rather damp day judging by the platform surfaces. The fact that it is carrying a 64B shedplate helps to date this image to the period prior to April 1961 or to a short period between May and December 1961 when the locomotive was allocated to Haymarket depot. It was subsequently withdrawn from St. Margaret's in October 1963. In many ways the A3s were unsuited to the Waverley route, for with their large driving wheels and three cylinders they were much more at home on long stretches of 80mph running hauling heavy expresses. The prime express on the Waverley route was the eponymous train which typically consisted of only 8 or 9 coaches. In addition the Waverley route was limited to a 70 mph maximum with many tight curves and much hill climbing all contributing to much lower speeds than the Pacific's former 'racing ground' of the ECML..*(Neville Stead Collection)*

WAVERLEY

86

Carlisle - Hawick - Galashiels - Edinburgh

Top left. On the 3rd. September 1955 Class A3 No. 60035 *Windsor Lad*, the first of the class numerically speaking, runs in with the 2.35pm Edinburgh to Carlisle service. Awaiting its arrival on the up platform is a barrow load of mailbags and on the left can be seen the substantial goods shed. The Pacific is carrying a 64B Haymarket shedplate, it being allocated here between April 1937 and April 1961 and again from August 1961 just prior to withdrawal the following month. Incidentally the champion racehorse 'Windsor Lad' was owned by the Maharajah of Rajpipla of the Indian state of Gujarat. *(W. A. C. Smith)*

Bottom left. Seen previously at St. Boswells this is a further view of Class D34 No. 62471 *Glen Falloch* operating the BLS 'Scott Country' railtour of 4 April 1959 photographed here before setting off from Galashiels. By this date there were still 25 examples of the class operating on BR but they had all gone by 1961. No. 62469 *Glen Douglas* was withdrawn from regular service in 1959 and was earmarked for preservation. It was repainted in the distinctive North British bronze green livery, assuming its former number of 256, and hauled a number of special trains until withdrawn in 1965 following which it was presented to Glasgow Corporation for static display in their Transport Museum. It can now be found in the city's Riverside Museum currently positioned next to a Glasgow tram. *(W. A. C. Smith)*

Above. Into this snow covered scene comes St. Margaret's based Class B1 No. 61191 with an up service. The signalman in his box is no doubt glad to be inside and probably has the stove going full blast. At one time there were no less than 5 signalboxes controlling the layout at Galashiels but in 1937 the LNER replaced them with this box at the end of the down platform. Surviving the closure of the Waverley Route to through traffic by 4 months it did not cease operations until April 1969 when the remaining goods only section from Edinburgh south to Hawick was cut back to Lady Victoria Pit near Newtongrange. *(Norris Forrest)*

Above. Another B1 in the shape of No. 61308 arrives at the up platform in more clement weather with a service from Edinburgh. This Thompson 4-6-0 had spent three spells at St. Margaret's depot during the late 1950s and early 1960s and the class were often to be found on locals between the Scottish capital and Galashiels sometimes operating via the Peebles loop. There was a good service between these two points with, in 1961 for example, some 20 departures from Edinburgh daily to Galashiels half of which operated via Peebles using DMUs. *(Neville Stead Collection)*

Top right. There was a small two road locoshed at Galashiels situated to the south of the station and in this image taken on 23rd. May 1961 Class J37 No. 64599, allocated to St. Margaret's depot in Edinburgh, is seen on shed. Whilst the depot here, which was a sub shed of 64A, would close in April 1962 it was subsequently used for locomotive storage purposes with its most prestigious occupants being Class A4 No. 60026 *Miles Beevor* and No. 60034 *Lord Faringdon* which shared the facilities here with 350hp diesel shunters. 'The Railway Observer' reported that both Pacifics were in store until late December 1963 but that on 27 December No. 60034 was hauled to Hawick shed by Standard No. 78049. The following day this A4 departed tender first light engine under its own steam to St. Margaret's shed but it was ultimately destined for store at Bathgate. By February 1964 the remaining Pacific No. 60026 had also moved to Bathgate from where it was recalled to active duty at Aberdeen's Ferryhill shed in April working the 3 hr. expresses to Glasgow until withdrawal at the end of 1965. The J37 seen here would end its days at Dunfermline Upper MPD in October 1965. *(A. E. Bennett)*

Bottom right. Talking of A4s, we conclude our coverage of Galashiels with a couple of shots of these Pacifics on railtour duty. The 'Scottish Rambler No. 4' tour of 18 April 1965 had No. 60031 *Golden Plover* at its head carrying reporting number 1X50. Departing Glasgow Queen Street High Level the tour proceeded to Edinburgh and Polmont. It then traversed the Waverley route to Longtown whence it took the former Gretna branch to Mossband and Kingmoor reaching Carlisle from where it returned to Glasgow via Carstairs. It was booked for an hour's layover at Galashiels and it was during this time that the photographer obtained his shot. *(W. A. C. Smith)*

Carlisle - Hawick - Galashiels - Edinburgh

Above. A couple of months later classmate No. 60027 *Merlin* was at the head of the Scottish Locomotive Preservation Fund (SLPF) special of 5 June 1965. Originating from Waverley station with A3 Class No. 60052 *Prince Palatine* in charge it traversed the ECML to Newcastle thence via Hexham to Carlisle from where the A4 took over for the run back to Edinburgh over the Waverley route. Use of the A4 was a bonus for tour participants as the A3 had developed a hot box and had to be removed at Carlisle being substituted by No. 60027. *(W. A. C. Smith)*

Top right. Although not initially included as a candidate for re-opening Stow was one of the intermediate stations between Galashiels and Edinburgh that, as a result of local pressure, would see services resume with the advent of the Borders Railway. In this 1955 image V3 Class No. 67669 waits at the station with a southbound local for Galashiels. The station handcart is a noteworthy feature seen to the right of the footbridge. *(Neville Stead Collection)*

Bottom right. Class B1 No. 61307 powering the 12:52 (SO) Edinburgh to Hawick service on 15 November 1958 is seen here at Fountainhall. The large number of enthusiasts present on the platform is explained by the running of a special train that same day by the Branch Line Society over the Lauder Light Railway which connected with the Waverley route here at Fountainhall. Standard Class 2 No. 78049 was in charge of the special which made a return trip to Lauder, regular passenger services to which had been withdrawn back in 1932. With goods traffic in decline and following a review of the Ministry of Food buffer depot which had been opened at Lauder during WW2, all freight traffic had finished at the end of September 1958 thus making this special the final train to run over the branch. *(W. A. C. Smith)*

STOW

FOUNTAINHALL

HERIOT

Above. On 11 May 1963 an unidentified D53XX class Bo-Bo diesel accelerates through Heriot station which at that time saw only a minimal service of three trains a day in each direction with southbound departures at 7.36am, 1.39pm and 4.58pm whilst those for Edinburgh left at 7.57am, 9.16am and 1.09pm. The station was unusual in that the platforms were staggered either side of the level crossing with the main station building, complete with drinking fountain, on the left being at ground level. *(W. A. C. Smith)*

Right. Judging by the two intending passengers walking along the up platform the service, from which this image was taken, is due to stop at Heriot. The main A7 road ran close to the station and was no doubt a more convenient artery than the sparse service offered by the railway for the few inhabitants that lived locally. The train is just crossing the trailing point that led to the small goods yard. The station became unstaffed along with many others in March 1967 and did not re-open with the coming of the Borders Railway which saw the abolition of the level crossing, the original gates of which survived in situ well into the 21st century. The road crossing was replaced with a bridge to the south although a pedestrian underpass was provided on the site of the former level crossing. *(Henry Priestley)*

Carlisle - Hawick - Galashiels - Edinburgh

FALAHILL

Top left. At 880 ft. above sea level Falahill was the summit of the northern half of the line between Hawick and Edinburgh. In this 1960 view V2 No. 60959 is slogging up the incline with a goods service seemingly without banking assistance which was often required. Spending periods at both Haymarket and St. Margaret's sheds in 1960 this 2-6-2 would be withdrawn from Aberdeen Ferryhill in July 1963. *(Henry Priestley)*

Bottom left. On 11 May 1963 St. Margaret's based Class B1 No. 61351 passes Falahill signalbox with the noon (SO) stopping service from Hawick to Edinburgh which called at all stations except Tynehead arriving at Waverley station at 1.49pm. Loops and sidings were provided at Falahill to accommodate banking locomotives and adjacent to the signalbox is the substantial water tower. *(W. A. C. Smith)*

Above. A strong crosswind is snatching away the locomotive's exhaust as Class B1 No. 61324 approaches Falahill on 11 May 1963 with a passenger service. Based at St. Margaret's depot this 4-6-0 would be withdrawn in October 1965. Falahill summit is reputed to be the tenth highest railway summit in the UK being only 35 feet lower than the more well known Shap summit on the WCML. *(W. A. C. Smith)*

WAVERLEY

TYNEHEAD

Above. The very basic facilities provided at Tynehead are revealed in this image taken on 11 May 1963. Situated in a deep cutting, the main station building was at road level with steeply sloping pathways down to the platforms. At the time of this view, just one train called each way on weekdays departing northbound at 8.03am (SX) and southbound at 4.48pm (SX). In light of this it is not too surprising that no new station has been provided here for the Borders Railway. Services to the goods yard, which was situated at a high level at the top of the cutting, were withdrawn in December 1964 with the station becoming unstaffed in March 1967. The former station building survives in residential use. *(W. A. C. Smith)*

Left. With Falahill box in the distance on the right hand side of this June 1960 view one of 64A's stock of Class V2s No. 60840 is on the short level section of track before the descent to Heriot with an up goods service which includes in its rake a couple of cattle vans. V2s were often to be seen seen in later years on the daily Halewood (Liverpool) to Bathgate freight trains conveying Ford cars on carflats. Due to the heavy loads carried booked motive power over the Waverley route was often a V2 double heading with a Black 5. *(Neville Stead Collection)*

WAVERLEY

FUSHIEBRIDGE

Both left. The first of two views of track renewal taking place adjacent to Tynehead signalbox. On this occasion the engineers train was hauled by a Clayton Type 1 (later Class 17) diesel. Notice in the first image that access to the goods yard line has been severed following withdrawal of facilities to the yard. In the second view a steam crane is lifting some replacement rails into position. Although these images are undated they are likely to have been taken on a Sunday when traffic over this section of line was light consisting of just three services in each direction.
(Norris Forrest)

Above. On 26 September 1964 a Type 2 BRCW diesel passes the site of the long closed Fushiebridge station with a down stopping service. October 1943 saw the end of passenger services here, it being half a mile south of the next station at Gorebridge, although it continued to serve as a halt for workmen at the nearby Catcune Mills for some years. The goods yard closed in January 1959 and there was at one time a siding serving Vogrie Colliery which led off behind the platelayers' hut seen on the left but this closed in 1938. The signalbox, which replaced an older box situated by the road bridge at the north end of the station, was constructed 300 yards to the south of the station in the early years of the 20th century to control access to the private sidings and the goods yard and it can be seen in the distance to the right of the coaching stock. *(W. A. C. Smith)*

GOREBRIDGE

Top left. Standard Class 4 No. 76049 gets away from Gorebridge's staggered platforms with the 12:00 (SO) service from Hawick to Edinburgh on 26 September 1964 whilst a porter on the up platform sorts some parcels in front of the signalbox. At this date the 2-6-0 was a Hawick based locomotive. Although houses were built on the trackbed after closure these were demolished to make way for the Borders Railway which again serves the settlement of Gorebridge which at peak times enjoys a half hourly service to and from the capital. *(W. A. C. Smith)*

Bottom left. Within easy commuting distance of Edinburgh, Gorebridge always provided a regular clientele for the railway. Here Standard tank No. 80055 hurries along near the town with the 5.11pm departure from Edinburgh on 11 May 1963 no doubt carrying home-going commuters and shoppers. Delivered new to Polmadie depot in December 1954 this 2-6-4T transferred to St. Margaret's in August 1962 from where it was withdrawn in September 1966 being cut up at the Faslane site of Shipbreaking Industries Ltd. in the following month. *(W. A. C. Smith)*

Above. Also on 11 May 1963 BRCW D5303 is seen calling at Gorebridge with an up service. After the end of regular steam working a variety of diesel locomotives handled passenger services especially Sulzer engined types (later Classes 24 and 26) and even Claytons (Class 17) were sometimes seen on local stopping services. Long distance trains were usually in the hands of Peaks (Class 45). Arriving in April 1960 D5303 spent a number of years based at Haymarket depot but a few months after this view was taken, in September 1963 to be exact, it suffered a severe fire when sparks from brake blocks ignited oil impregnated waste. It was repaired however and lasted in service for a creditable 35 years until withdrawal in October 1993. The truncated goods siding seen to the left containing the solitary van parked by the buffer stop was formerly connected to both the up and down mainlines at its southern end. *(W. A. C. Smith)*

NEWTONGRANGE

Above. Type 2 D5308 diesel heads south on 11 May 1963 passing the complex on the right which comprised Lady Victoria Pit situated between Gorebridge and Newtongrange. The signalbox controlled access to the exchange sidings seen on the right serving the colliery whilst spoil heaps of colliery waste are evident to the left of the main line. Opening in 1895 as Newbattle Colliery, Scotland's first 'super pit', it had its own fleet of small industrial tank locomotives including designs by Andrew Barclay and Grant Ritchie & Co. Taken over by the NCB upon nationalisation in 1947 coal production ceased in 1981 although it survived to become the Scottish Mining Museum created in 1984 as one the best preserved Victorian collieries in Europe with over 100,000 artefacts in its collection. *(W. A. C. Smith)*

Top right. Class J37 No. 64552 is seen at Lady Victoria pit sidings in this image from the early 1960s. Constructed in 1916 this veteran 0-6-0 would last in service until withdrawal from Dunfermline Upper (62C) shed in October 1964. *(Norris Forrest)*

Bottom right. Standard Class 4 No. 76049 seen previously at Gorebridge on 26 September 1964 also had charge of the 12 noon service from Hawick three weeks earlier on 5 September seen here about to pass under the footbridge at Newtongrange. Situated in a cutting passengers accessed the platforms via the Booking Office and steps down from the footbridge. *(James L. Stevenson)*

Carlisle - Hawick - Galashiels - Edinburgh

WAVERLEY

Carlisle - Hawick - Galashiels - Edinburgh

Left. D5315 powers non stop through Newtongrange with the 5.52pm service from Edinburgh to Carlisle on 24 May 1962, the first stop being at Stow thence principal stations only until arrival into Carlisle at 8.47pm. The station had become unstaffed eighteen months earlier in December 1960 and today a new station has been provided for the Borders Railway situated to the south of the site of the former station. *(W. A. C. Smith)*

Below. Class B1 No. 61341 has charge of the 5.11pm Edinburgh Waverley – Galashiels stopping service and is seen here crossing Newbattle Viaduct over River South Esk on 24 May 1962. If running to time it was scheduled to arrive in Galashiels at 6.20pm. Fortunately being a Grade B listed structure the 23 arch Newbattle Viaduct had not been demolished and was therefore available for use by the Borders Railway. This St. Margaret's based 4-6-0 would be withdrawn at the end of the following year being scrapped at Cowlairs Works. *(W. A. C. Smith)*

HARDENGREEN JUNCTION

ESKBANK & DALKEITH

MILLERHILL

Top left. Passing Hardengreen's lofty signalbox, which controlled the junction for the Peebles loop to Galashiels, one of the Clayton/Beyer Peacock Bo-Bo diesel locomotives No. D8615 proceeds southwards on 23rd. August 1966. Constructed in March 1965 this Type 1, later Class 17/3, had been allocated to Haymarket depot in May 1966 and was to give, like many of this short lived class which was considered even less successful than the ill fated Metrovick diesels, just 6½ years service before withdrawal in October 1971 and subsequent scrapping at St. Rollox Works. The twin Paxman engines with which they were fitted proved unreliable being susceptible to camshaft and cylinder head problems and the class achieved an overall availability of barely 60% even following considerable modifications. Forward visibility, which had dictated the centre cab design of this type, was not good with crews struggling to see the area immediately in front of the locomotive. The working depicted here would seem to indicate an engineers' inspection with a solitary clerestory saloon being towed behind the locomotive. *(Leslie Sandler)*

Bottom left. Class V3 2-6-2T No. 67668 running bunker first departs southwards from Eskbank & Dalkeith station with a three coach local for Galashiels. Notice the strategically positioned large advertising hoarding on the left designed to catch the eye of the captive market of passengers ensconced within passing coaching stock. A new repositioned station serves the current Borders Railway. *(W. A. C. Smith)*

Above. Class A3 No. 60093 *Coronach* heads the 5.52 pm Edinburgh Waverley – Carlisle service through the middle of Millerhill yard on 3rd. April 1961. This extensive marshalling yard, begun in 1958, was still under construction at this time with the Up yard opening in 1962 and the Down yard the following year. Despite being initially successful like many such yards Millerhill suffered declining goods traffic such that the Down yard closed only 20 years later in 1983 with the Up yard severely truncated to provide for the few remaining freight and engineers' trains. *(W. A. C. Smith)*

Above. Standard Class 3 2-6-0 No. 77009, allocated to Grangemouth depot, makes its way past Millerhill yard with a lengthy rake of empty mineral wagons in this 1964 view. A diesel shunter is dealing with a train of Palvans on the right whilst in the distance a Type 2 Bo-Bo locomotive is held at signals. *(Neville Stead Collection)*

Top right. A3 Pacific No. 60035 *Windsor Lad* puts on the power with a southbound service as it accelerates through Portobello station on the avoiding line on an unrecorded date in 1960. Portobello East Junction was the point where the Waverley route joined the ECML. *(Neville Stead Collection)*

Bottom right. Proceeding rather more slowly is the 4.10pm Edinburgh to Hawick stopping train hauled by Class B1 No. 61029 *Chamois* on 21st. August 1964 which was scheduled to take 6 minutes to reach this popular seaside resort from Waverley. This was one of four Waverley line services that called at Portobello station on weekdays at this time but the following month the station would close with the withdrawal of local services from Musselburgh to Edinburgh. *(W. A. C. Smith)*

Carlisle - Hawick - Galashiels - Edinburgh

PORTOBELLO

Above. A similar scene but five years previously sees Class J37 No. 64562 depositing its passengers at Portobello whilst the fireman looks back along the train to catch sight of the guard's green flag prior to departure. Dating from 1918 this 0-6-0 would go on to survive until withdrawal in November 1963. The 365 foot chimney of Portobello power station, which was claimed to be the most efficient in the UK in the early 1950s, dominates the skyline to the right of this 1959 image. In 1936 the art deco-style Portobello Bathing Pool was built adjacent to the power station and whilst this may seem to be a strange juxtaposition, heated waste water from the power station was piped underneath the pool to warm it when required. The power station was decommissioned in the 1970s. *(Neville Stead Collection)*

Top right. And so we come to 'Auld Reekie', the somewhat uncomplimentary name given to the city of Edinburgh in the 17th century when smoke from open coal and peat fires hung smog-like over the city. In this panorama of the eastern approaches to Waverley station taken on 13 July 1963 Class B1 No. 61029 *Chamois* running tender first passes the substantial East signalbox whilst shunting carriage stock. Also in view are a couple of diesel shunters, one in the station precincts with another in the goods yard seen to the left. An English Electric Type 4 locomotive is apparent on the far right whilst steam from an unidentified tank locomotive is visible to the right of the signalbox. The corner of New Street Bus garage can just be made out to the left of the signalbox and a fine procession of Edinburgh Corporation (subsequently renamed Lothian Region Transport in 1975) double and single deckers can be seen crossing North Bridge in the background. Also of interest is what appears to be a small fleet of Morris J Type vans parked on Calton Road in the bottom right hand corner of this image. These 10cwt vans were popular with a number of high street names such as John Lewis, Dunlop, the RAC and Walls ice cream. *(S. Rickard)*

Bottom right. This 1954 view shows, at the west end of the station, Class A2 No. 60509 appropriately named *Waverley*. This Haymarket based locomotive, one of the sub class designated A2/1, would be withdrawn in August 1960. In the background is the magnificent North British hotel which after extensive refurbishment in 1988 changed its name to 'The Balmoral'. Although initially named after the local railway company, the NBR, this was apparently not popular with some Scots at the time being considered demeaning to Scotland's status as a separate nation within the United Kingdom. *(Neville Stead Collection)*

EDINBURGH WAVERLEY

WAVERLEY

Right. This Gloucester RCW twin set DMU, leading coach numbered SC56310, makes a substantial contribution to the pollution within the confines of Waverley station whilst waiting departure time with a Galashiels via Peebles service. As previously noted Peebles trains used the Waverley route as far as Hardengreen Junction with some continuing on to Galashiels. DMUs were introduced to regular service on this the route as early as 1958 being one of the first lines in the country to benefit from their inherently lower costs and improved journey times. The unit pictured here was delivered new to Leith Central depot at the end of 1957 and would remain in service until withdrawal from Norwich in October 1972 being subsequently scrapped at Cohens of Kettering. Dieselisation did not however save the Peebles route from closure which came in February 1962 rendering Peebles-shire the distinction of being the first mainland Scottish county to lose all its railway services. *(George C. Bett)*

Below. Our final view of Scotland's capital city station is this image dating from 21st. September 1965 taken from Princes Street gardens and illustrating the sunken nature of the railway hereabouts with North Bridge in the background and Waverley Bridge in the foreground spanning the station. Steam still lingered at this date but diesels are very much in evidence with two DMUs, a shunter and a Class 40 to the fore. *(David Anderson Collection)*